The Story of

ASTRONOMY IN EDINBURGH

from its beginnings until 1975

to my wife

Small magellanic cloud taken with 48-inch Schmidt telescope in Australia.

The Story of

ASTRONOMY IN EDINBURGH

from its beginnings until 1975

HERMANN A. BRÜCK
Professor Emeritus in the University of Edinburgh
Formerly Astronomer Royal for Scotland

EDINBURGH
1983

Produced by
Edinburgh University Press
22 George Square
Edinburgh

ISBN 0 85224 480 0

Printed in Great Britain by
Alna Press
Broxburn, West Lothian

CONTENTS

PREFACE

ON THE occasion of the 150th Anniversary of the foundation of the Royal Observatory Edinburgh in 1972 I wrote a little booklet on the history of the Observatory and of the Chair of Astronomy in the University of Edinburgh. In that booklet I gave a brief account of astronomical activities in Edinburgh before 1972. The more I enquired in writing that booklet into the lives and achievements of my predecessors the more interesting many of them seemed to be, and I decided to give a somewhat fuller account of their work when I had more time.

The 400th Anniversary of the University of Edinburgh seemed a very suitable occasion for that history. In preparing it I realised that twenty-five years had elapsed since my own appointment to Edinburgh and that my own time here was already slipping into past history. I decided therefore to add to the story of my predecessors an account of the activities of the Royal Observatory and of the Astronomy Department of the University between the years of my appointment in 1957 and my retirement in 1975.

These years have been a time of unequalled expansion and I have therefore not been able to describe these recent scientific activities in the same detail in which I could record the work under my predecessors. This has also meant that, much to my regret, I could not mention explicitly everyone of the dozens of dedicated and enthusiastic members of staff who worked with me during hectic years and who helped to put the Observatory into a prominent position within the international astronomical community.

Complete lists of all Observatory and Departmental staff can be found in the Annual Reports which I addressed first to the Secretary of State and later to the Science Research Council. The same Reports give full accounts of all scientific activities in the Observatory and University Department. Shorter versions of Annual Reports appeared in the *Monthly Notices* and later in the *Quarterly Journal of the Royal Astronomical Society*.

The Bibliography includes titles of publications which have been issued by the Observatory from its foundation until 1975. The *Astronomical Observations* were started by Henderson in 1838 and continued by Piazzi Smyth, the *Annals* were started by Copeland in 1902 and continued by Dyson, and the *Publications* were started by Greaves in 1938 who also instituted the *Communications*. Both the *Publications* and the *Communications* were continued by myself.

In writing this book I have been very fortunate in being assured at all times of the friendly interest of many colleagues. I am particularly grateful to Professor Malcolm S. Longair, Astronomer Royal for Scotland, for the way in which he has placed all the resources of the Royal Observatory, which is now an establishment of the Science and Engineering Research Council, at my disposal. I also wish to express my thanks to Professor Vincent C. Reddish and many members of the Observatory staff for the help they have given me. I have greatly appreciated the assistance of the Observatory's librarian, Mr A. R. Macdonald, and of Mrs M. F. I. Smyth while she was in charge of the Crawford Collection and the Observatory archives. Invaluable has been the ever ready help of Mr B. McInnes, now the Observatory's Secretary, for the tracing of records and for the supply of photographs of the Observatory's activities in the fields of satellite tracking and site testing. My thanks are extended to Mr B. W. Hadley and his team in the Observatory's Photolabs for the splendid pictures of portraits, instruments and other items important in the history of the Royal Observatory. In earlier years much of the general photographic work in the Observatory was carried out by Mr A. McLachlan, whose skill is evident in various photographs in this book ranging from scientific instruments to social occasions.

I am also happy to be able to thank Mr Allen Simpson and the Photographic Department of the Royal Scottish Museum for the photographs of one of James Short's early telescopes and of two clocks of historical interest. The photographs of UK instruments overseas and the UK Schmidt Telescope astronomical photographs have been supplied by Photolabs, Royal Observatory Edinburgh, and are reproduced by courtesy of the Science and Engineering Research Council. Photographs of the Royal Visit, of Sir Edward Appleton and Lady Tweedsmuir are reproduced by permission of The Scotsman Publications Ltd.

The photograph of the bust of Robert Blair is reproduced by permission of the University of Edinburgh, and that of the portrait of Charles Piazzi Smyth by permission of the Royal Society of Edinburgh.

I owe a special debt of gratitude to the Carnegie Trust for the Universities of Scotland and to the Royal Observatory (Edinburgh) Trust for grants which have made it possible to start with the printing of this book at a time of great financial stringency.

My warmest thanks are due to the staff of the Edinburgh University Press for the great care and patience with which they have prepared this book.

This book is seen as a modest contribution to the celebrations of the 400th Anniversary of the University of Edinburgh.

H. A. BRÜCK
January 1983

Part One

The Town, the University and Calton Hill
1610–1888

1

EARLY DAYS AND THE ASTRONOMICAL INSTITUTION OF EDINBURGH

AN INTEREST in astronomy as an academic discipline started in Edinburgh with the opening in October 1583 of the Town's College, the present University of Edinburgh. In the early days of the College astronomy was taught as part of mathematics in a broad four-year course in Arts which was given not by specialist Professors, but by the 'Regents of Philosophy' each of whom was responsible for the teaching of every subject in the curriculum. The standards reached in the teaching of astronomy can be seen from the 'Theses Sphaericae' and 'Theses Astronomicae' for candidates for the MA degree which are preserved in the University Library. Other University records speak of early astronomical observations in the College like that of a comet in 1618, one of three which played some role in Kepler's well-known treatise of the following year in which he stressed his view of the celestial nature of comets against the then widely held opinion that comets are purely atmospheric phenomena.

The 17th century saw the foundation of the first specialist Chairs in the College of Edinburgh, one of the earliest being a Chair of Mathematics which was established in 1620. Great expectations for the future of astronomy in Edinburgh were roused when in 1674 James Gregory was appointed to this Chair in succession to Thomas Crauford. The first of the family of the 'academic Gregories', three of whom were to follow each other in the same Edinburgh Chair of Mathematics, James, a graduate of Marischal College, Aberdeen, was a mathematician of the highest reputation. Apart from publications in pure mathematics he was responsible for a major work in optics, his 'Optica Promota' which he published in 1663 at the age of twenty-four and in which he gave a detailed description of a reflecting telescope for which he claimed parity with if not priority over Isaac Newton.

James Gregory came to Edinburgh from St Andrews where for five years he had held a Chair of Mathematics which had been established in that University by Charles II. It was a major tragedy for Edinburgh astronomy as well as for mathematics when Gregory's move was followed only a year later by his early death at the age of thirty-seven. It is said that Gregory was suddenly attacked by 'blindness' while showing the satellites of Jupiter to a group of students through one of his telescopes.

The next fifty years saw little of astronomical activity in Edinburgh which had to wait until the appointment to the Chair of Mathematics in 1725 of the great Colin Maclaurin, one of Scotland's most distinguished mathematicians. At the time of his appointment Maclaurin was a young man of only

THESES ASTRONOMICÆ.

3. *Oculus igitur quò propiùs ad globum accedit, eò minorem videt ipſius partem: & quò longiùs, majorem.*

4. *Lunæ in Apogæo major pars videtur quàm in Perigæo.*

IX. Quæ ſub majoribus angulis ſpectantur, majora apparēt: quæ ſub minoribus, minora: ut Optici demōſtrant.

1. *Videmur itaque majorem globi partem intueri, quò propiùs: & minorem, quò longiùs intuemur.*

2. *Ideò Luna in Perigæo maior videtur quàm in Apogæo.*

3. *Hinc fit, ut Luna licet omnium ſtellarum minima, excepto Mercurio, major omnibus videatur: Soli verò æqualis, qui tamen infinitis ferè partibus major eſt.*

X. Lunæ à terra diſtantia maxima putatur eſſe diametrorum terræ $32\frac{1}{11}$: minima verò, tantùm $16\frac{1}{2}$.

1. *Diſtantia Lunæ maxima ad ejuſdē minimam dupla eſt.*

2. *Cùm Luna non videatur duplò major in Perigæo quàm in Auge, argumentum Copernici paralogiſmus eſt, quò cōcludit duplò majorē videri magnitudinem, quæ duplò propinquior eſt.*

3. *Diſtantiæ Ptolemæi poſſunt eſſe veræ, dum interim diametri in utroque intervallo plurimùm non videantur differre.*

4. *Ac proinde Optica dictabit æquales magnitudines ab oculo inæqualiter diſtantes, habere minorem rationem angulorum ſub quibus cernuntur, quàm diſtantiarum: nam anguli ſenſibiliter non differentes, diſtantiam valdè ſenſibilem habere poſſunt.*

FINIS.

THESES SPHÆRICÆ.

5. Neque mentes abſtractæ quamvis ſola mente noſcantur.

6. Ergo non movent mentem vel proprio, vel accidentium ipſis inhærentium phantaſmate, ſed phantaſmatis operum ſenſilium.

39. Animalium facultatibus judicantibus varia ſeſe offerunt, vel vitanda, vel amplectenda, quibus animalia ad motum invitantur: quæ tamen non moventur niſi mens vel Phantaſia excitet appetitum, & moveat organa ſine quibus non fit motus.

1. Quare in efficiendo motu inceſſus, primo objectum movet facultatem noſcentem, mox facultas formata ab objecto judicato, tanquam utili vel adverſo, excitat appetitum ad proſequutionem vel fugam, quam ſequitur conſpicua corporis motio.

2. Adeo ut ad hunc motum quatuor conſpirent principia, excitantia, dirigentia, exequentia, famulantia.

30. Quod ex ſe ipſo movetur, nunquam deſinit moveri ex eo quod aliud quiddam motum ſiſtat. 7. Phyſ. cap. 1.

1. Neceſſe eſt ſi quid moveri deſinit eo quod aliud quiddam ſiſtatur, hoc ab altero moveri. Ibid.

2. Id totum quod quieſcente parte quieſcit, non movetur per ſe, nec eſt primum ſe movens.

3. Cum animal quieſcente corpore, vel ſpiritibus, non moveatur, non erit primum ſe movens abſolute, ſed per partem movebitur.

4. Ergo cum dicimus animal movere ſeipſum per ſe & primo, ſumitur primum pro integro & toto, id eſt, non per aliud, non enim movetur per aliud totum prius.

41. Motor cœli cum ſit efficiens, movet a ſe quaſi pellendo, & cum ſit finis, ad ſe movet quaſi trahendo: lapis ut a proprio ſuo loco trahitur, ſic a ſubſequente aere pellitur: animal dum movetur ſupra terram, partim pellit a terra, partim trahit, & partes vel a cardine oſſium pelluntur, vel ad eum trahuntur.

1. Ergo non tantum motus violenti, vectio, vertigo, & motus animalis, ſed omnis omnino motus fit tractione & pulſione.

2. Et omnis omnino motus requirit aliquod quieſcens.

Theſes Sphæricæ.

1. CVm ſtella moveatur orbi infixa, & unius ſimplicis unus ſit motus, ac Sphæra ſuperior moveat inferiorē, non contra:

1. Septem Planetæ non in una, ſed in ſeptem Sphæris ferentur, & ſtellæ fixæ in octava.

2. Cum ergo tres motus ſint trium, erunt tres Sphæræ Planetis ſuperiores.

E.

Theses Astronomicae. Examination questions on Astronomy from the 17th century in the University of Edinburgh (Edinburgh University Library).

James Short, Edinburgh telescope maker (Royal Observatory Edinburgh).

twenty-seven, full of enthusiasm and keen on lecturing. His lectures included a very popular course on 'Experimental Philosophy' which he illustrated by practical demonstrations including observations of the Moon, planets and stars through telescopes which he had mounted on the roof of that part of the University's old College which a hundred years later was to be converted into William Playfair's Upper Library.

One of Maclaurin's ambitions was the establishment in Edinburgh of a properly equipped astronomical observatory whose instruments might be available both to University students and to amateur astronomers of the general public. In this aim he had the strong support of George Drummond, the Lord Provost and perhaps the most influential citizen of Edinburgh in the 18th century. Other supporters came from the circle around the Earl of Morton and Sir John Clerk of Penicuik, and a formal proposal for the foundation of an Edinburgh Observatory was in fact put forward in 1736, but the times were inauspicious. The same year saw the 'Porteous Riots' in which Edinburgh became the scene of grave disturbances leading to the lynching of Captain Porteous of the Town Guard. Undaunted, Maclaurin and the Earl of Morton pursued the matter once again four years later when they set up a special observatory fund for which Maclaurin offered the proceeds of his public lectures, but nothing was actually achieved, and the Jacobite rising in 1745 and Maclaurin's untimely death a year later put an end to this particular endeavour.

Maclaurin's efforts on behalf of astronomy were not entirely wasted, however. His lectures aroused the particular enthusiasm of James Short, who had come to Edinburgh to be trained for the Ministry of the Church of Scotland, but under the influence of Maclaurin decided to devote himself instead to the pursuit of science. He became particularly interested in the field of optics, both practical and theoretical, and with the support and guidance of Maclaurin Short started in the early 1730s constructing reflecting telescopes of the Gregorian design. The mirrors of his telescopes were made of speculum and in the course of time were manufactured to such precision that Maclaurin was able to describe them as 'by far the best of their lengths that have yet been executed'.

While he was working in Edinburgh James Short joined the 'Society for improving Arts and Sciences, especially Natural Knowledge' which was founded in 1737 with the Earl of Morton as President, Sir John Clerk as Vice-President and Professor Maclaurin as Secretary. Two years later Short moved to London where he became the leading telescope maker of the day, some thirty years before William Herschel began his work. Short's instruments ranged from small portables to major telescopes with apertures as large as 18 inches. For a long time Short's telescopes could be found in many parts of the world, both in observatories and private houses. When James Short died in 1768 his business was taken over by his younger brother Thomas who was an optician and instrument maker like James and who had started work in Edinburgh's port of Leith. In 1770 Thomas Short returned from London to Edinburgh bringing with him parts of a 12-inch telescope which had originally been ordered by the King of Denmark, but which he now intended to use in a private observatory of his own which he wanted to run as a commercial undertaking.

Having first placed his telescope on the roof of Heriot's Hospital, Thomas Short soon applied to Edinburgh Town Council for the lease of a plot of ground on the Calton Hill where he expected to be able to establish his observatory to greater advantage – and herewith begins the next chapter in the story of an astronomical observatory in Edinburgh. When Short's plans became known it appeared to the Trustees of the Observatory Fund which had been raised earlier by the Earl of Morton and Professor Maclaurin and which then amounted to £400 that Short's private observatory might well become the Edinburgh Observatory envisaged by the City and the University if only some agreement could be reached between the interested parties. This was in fact done in the sense that the Trustees placed their Fund at Short's disposal and the City provided an observatory site on Calton Hill both on condition that the observatory would be open to students of Edinburgh University on special terms and that on Short's death any observatory buildings and instruments would become the property of the City of Edinburgh.

Following the signing of this agreement the foundation stone of this first Edinburgh observatory was laid on the Calton Hill in the summer of 1776 by Lord Provost James Stodart in the presence of many members of the Town Council and of the Senatus of the University. Things seemed to move at last in a promising direction and all might have been well if plans for the observatory building had been a little less ambitious and expensive. The plans for the observatory had been drawn up by James Craig, the young architect who gave Edinburgh's New Town its original shape and form. It

Top: Gregorian telescope by James Short made in Edinburgh in 1735 (Royal Scottish Museum). *Centre:* The Old Observatory on Calton Hill 1792. *Bottom:* Playfair's New Observatory on Calton Hill 1824.

was based on the idea – it is said upon the passing suggestion of Robert Adam – that the chosen site on the Calton Hill called for a building in the style of a fortification surrounded by a wall with Gothic towers at its corners.

It was soon found that the money collected for the building of the observatory was exhausted after the construction of only one of the Gothic towers – the one which Thomas Short had chosen as his residence. Hugo Arnot in his *History of Edinburgh* reports that the observatory was still unfinished when Short died in 1788 and when according to the terms of the agreement of 1776 the building became the property of the City of Edinburgh. It appears that for a time the city's rights were contested by members of the Short family who tried to gain possession of the property by force and had to be evicted by the Town Guard.

The building of the 'Old Observatory' as it came to be called later was finally completed by the City in 1792, sixteen years after its foundation stone had been laid in an atmosphere of the highest expectations. Its final structure was on a very modest scale and since there were no funds for the acquisition of instruments this first Edinburgh observatory was never used for any serious astronomical work. The situation was all the more unfortunate as in the meantime a Chair of Practical Astronomy had been established in the University of Edinburgh.

The Observatory cause was taken up again, and at last successfully, in the beginning of the 19th century by members of the 'Astronomical Institution of Edinburgh'. This was a Society of private citizens which was founded in 1811 and, preceding the foundation of the Royal Astronomical Society in London by nine years, was in fact the first British Society whose activities were entirely devoted to the pursuit of astronomy. It counted amongst its members many of the leading Edinburgh citizens of the day. Its first President was John Playfair who having been Professor of Mathematics for twenty years moved to the Edinburgh Chair of Natural Philosophy in 1805. He was a man of immense erudition whose interest in geology led to his well-known *Illustrations of the Huttonian Theory of the Earth* which he published in 1802 and which had a profound influence on the advancement of geology.

A major aim of the Astronomical Institution was the establishment – long overdue as members felt – of a proper observatory which could be used both for scientific research and for popular observation. The intention was to raise the funds by private subscriptions for the foundation of a 'scientific' observatory where 'accurate observations could be made with the best procurable instruments' and also of a 'popular observatory and physical cabinet' where members could make 'observations and experiments in astronomy or any other science'.

Following the examination of a number of possible sites in and around Edinburgh the members of the Institution decided to accept Thomas Short's choice of the Calton Hill as the most suitable site for their two observatories. They applied to the Town Council for the lease of the old Gothic Tower to serve as their popular observatory and at the same time for the lease of some adjoining ground on which they could build their new scientific establishment. Their application was granted on condition that the Tower and any other new observatory building should revert to the City of Edinburgh if they ever ceased to be used for astronomical purposes.

Having mounted a few minor instruments in the old Gothic Tower the members of the Institution started the planning of their scientific observatory. A design for its building was put forward by William Playfair, the young nephew of Professor John Playfair whom he had joined in Edinburgh after the death of his father in 1793. William Playfair having subsequently worked in London and visited parts of France returned in 1816 to Scotland at the age of twenty-seven when he became responsible for the design of many of Edinburgh's most prominent classical buildings.

Playfair's plan for the observatory though less strange than Craig's had been for the original unfinished building was still on distinctly unusual lines. Modelled on the Greek Temple of the Winds the observatory was envisaged by Playfair as a cruciform Roman Doric structure with six pillars in front of each of its four strictly equal sides and with a prominent dome for a telescope in its centre. The architecture was not dissimilar to Palladio's well-known Villa Rotunda in Vicenza.

Playfair's ideas unusual as they were for an observatory were accepted by the members of the Astronomical Institution, and the foundation stone of this 'New Observatory' was laid a little to the east of the old Gothic Tower on 25th April 1818 by Sir George Mackenzie who as Vice-President of the Institution represented Professor Playfair who was too ill to attend. The object of the observatory was classically expressed at the time on an engraved platinum plate saying:

> 'Ne diutius urbi clarissimae scientiam omnium pulcherrimam atque amplissimam excolendi facultas deesset' (that a renowned city should not any longer lack the facilities for the pursuit of the fairest and grandest of the sciences).

The building operations took six years so that Professor Playfair who died in 1819, a year after the laying of its foundation stone, never saw the completion of the new observatory for whose creation he had been so largely responsible. His efforts are recalled, however, in a monument which William Playfair erected close to the observatory as a memorial to his uncle.

The observatory is close to another well-known building designed by William Playfair whose site on the Calton Hill was influenced by objections raised to its originally proposed position by members of the Astronomical Institution. This is the 'National Monument' designed as a facsimile of the Parthenon in Athens and built as a great memorial to all those killed in the Napoleonic wars. Its construction was started two years after the completion of the observatory, but had to be halted three years later when the money collected by an appeal launched in 1822 and signed by Sir Walter Scott amongst others had been exhausted.

Lack of money presented a major problem also for the activities of the new observatory. The funds of the Astronomical Institution had been entirely used up by building costs leaving no money for the purchase of instruments. However, two years after the completion of the observatory building the Government accepted an appeal from the members of the Institution and came forward with a substantial special grant of £2000 to be used for the acquisition of equipment.

The instruments which were acquired out of this Government grant were

Left: William Wallace, Professor of Mathematics at Edinburgh University from 1819 to 1838 and observer for the Astronomical Institution on Calton Hill. The portrait was presented to the Royal Observatory by Wallace's daughters after the death of their brother Alexander, Assistant at the Observatory on Calton Hill from 1834 to 1880. *Right:* A 3-inch Gregorian telescope presented to the Astronomical Institution of Edinburgh in 1823 by Sir George Mackenzie, one of the founder members (Royal Observatory Edinburgh).

to constitute the main equipment of the observatory for more than fifty years. All of them came from well-known instrument makers of the time. The two principal instruments were an 8-foot Transit Circle by the firm of Repsold in Hamburg which had a 6-inch object glass made by Fraunhofer and Utzschneider in Munich, and a 6-foot Mural Circle by Troughton and Simms of London with a 4-inch object glass, an instrument very similar to a Mural Circle at the Royal Observatory in Greenwich. There was also a small Altitude and Azimuth Instrument constructed by Troughton and Simms. When these instruments were installed at the Observatory they were placed under the supervision of Professor William Wallace who from 1819 had been Professor of Mathematics in the University and who acted as 'Observer' for the Astronomical Institution.

In August 1822 King George IV paid his celebrated visit to Edinburgh, the first of a Hanoverian monarch to Scotland, amid scenes of public enthusiasm and jubilation. Contemplating how best to pay homage to His Majesty on that great occasion, the Astronomical Institution at its meeting the previous month had decided to present a loyal Address, and had prepared the text which read:

> May it please Your Majesty
> We, Your Majesty's loyal subjects the Astronomical Institution of Edinburgh beg leave to approach Your Majesty and to join in the universal acclamation of gratitude and joy which resounds throughout your ancient kingdom of Scotland on the happy occasion of Your Majesty's arrival.
>
> Associated for the advancement of that Science which directs the strength and the glory of Your Majesty's arms to the remote corners of the Earth and which enables your enterprising subjects to render your Empire the richest and mightiest in the World, we feel it our duty thus to offer to our Monarch the expression of our ardent desire that Your Majesty may long live to fill that Throne from which your august Patronage to everything connected with the extensions of useful knowledge is so graciously and so bountifully dispensed.

The Address was dated August 16 (the day after the King's arrival in Scotland) and signed by Sir George Mackenzie, President of the Institution. Sir George, who in his capacity as Vice-President of the Royal Company of Archers, the Monarch's Bodyguard for Scotland, was in attendance on the King during the visit, was able to have the Address presented in Edinburgh through the Secretary of State, Robert Peel. A few days later, on August 20, a reply was received:

> I have laid before the King the loyal and dutiful address of the Astronomical Institution of Edinburgh which I have the satisfaction of acquainting you His Majesty was pleased to receive very graciously. I have the pleasure to add that His Majesty is graciously pleased to permit the Observatory on the Calton Hill to be styled The Royal Observatory of King George the Fourth. I have the honour to be, Sir, your obedient and humble servant
>
> Robert Peel

The splendid new title nominally elevated Scotland's Observatory to the rank of England's great Royal Observatory at Greenwich. In practice it brought no immediate changes, though the way was paved for further representations in the years that followed for Government funds. The Royal Observatory at Greenwich, founded by King Charles II as early as 1675 was a publicly supported scientific institution of world renown under the direction of the Astronomer Royal. It was to be some years before the Edinburgh Observatory was to attain the same professional status as its sister establishment south of the border.

The specific name of George the Fourth in the Observatory's title appears to have been dropped after the death of that King in 1838. The Astronomical Institution's loyal address to his successor William IV on his accession refers simply to the Royal Observatory, and thereafter this or the Royal Observatory of Edinburgh became its official description.

2

ROBERT BLAIR
FIRST REGIUS PROFESSOR OF ASTRONOMY

IN THE meantime and long before the new observatory was actually established the University authorities had taken a major step forward in their support for astronomy with the creation as early as 1785 of a Chair of Practical Astronomy. Their initiative was to some extent conditioned by efforts of the Edinburgh Town Council which aimed at the provision of some practical instruction in navigation for members of the merchant service sailing from the City's Port of Leith.

The new Chair was to be a Regius one and its original practical aspect was expressed clearly in the Royal Warrant signed by George III which speaks of 'the great advantages which Navigation and the useful Arts derive from the cultivation of Practical Astronomy and that it is of great importance in the education of youth, and especially of those who are destined for the naval line, that they be instructed in the principles and practice of Astronomical Science'.

The University's move was widely welcomed, but their appointment of the first Professor, Robert Blair, unusual in itself, turned out to be less than fortunate in the course of time. Blair was the son of the Minister of Garvald in East Lothian where he was born in 1748. He studied Medicine at Edinburgh University before joining the Royal Navy in 1773 first as an apprentice to naval surgeons and later as a naval surgeon in his own right. Having spent ten years with the Navy in the Indies Blair returned to England in the early 1780s.

Blair's experiences at sea had roused his interest in navigation and navigational instruments. His first scientific publication which was later published as an Appendix to the Nautical Almanac for 1788 was concerned with the operation of the Quadrant and a proposal for an improved method

Robert Blair, first Regius Professor of Astronomy at the University. Bust in the Upper Library, University of Edinburgh.

for the adjustment of the instrument. The Quadrant which had been first described in 1731 by John Hadley in a paper to the Royal Society was the predecessor of the modern sextant and was employed in the same way for the measurement of angular distances. At sea the quadrant was used to determine altitudes of the Sun above the ship's horizon and so the geographical latitude of its position. Measurements of angular distances of stars or planets from the Moon could then provide the ship's longitude. Blair's proposal for an improvement of the quadrant brought him in 1783 a valuable prize of £100 from the Commissioners of Longitude.

In 1785 Blair, then aged thirty-seven, was awarded the degree of Doctor of Medicine by Edinburgh University and in September of the same year he was appointed to the new Chair of Practical Astronomy and was formally inducted into the University Senatus in February 1786. In the following year he started work on the improvement of the optical performance of refracting telescopes. Earlier efforts in the same direction had been made by John Dollond who produced the first 'achromatic' refracting telescope in 1758, but most telescopes of the 18th century had remained reflectors constructed along the lines which had been suggested by Gregory, Newton and Cassegrain. The use of mirrors had ensured that optical images did not suffer from the effect of 'chromatic aberration' which blurs images of white or multicoloured objects when produced by lenses.

Dollond in his refracting telescope had reduced the effect of chromatic aberration by the use of compound lenses made up of crown and flint glass, but flawless discs of flint of any size were difficult to make and remained indeed rare until the beginning of the 19th century. Blair's aim in his experiments was to find a replacement for the unsatisfactory flint by a fluid with similar optical properties. Combinations of glass and fluid lenses had first been suggested in 1747 in a purely theoretical paper by the great Swiss mathematician Leonhard Euler who proposed that suitable combinations could lead to a high degree of achromatism. With these ideas in mind Blair embarked on an extensive series of experiments for the determination of the refractive indices and dispersions of a range of fluids. To measure angles of refraction he used a method first proposed by Newton, but improving it by the use of a Hadley Quadrant. Blair described his experiments in considerable detail in a substantial paper which was read at two meetings of the Royal Society of Edinburgh in January and April 1791.

Dollond's compound lenses could only reduce, not eliminate the effect of chromatic aberration which showed up as a 'secondary spectrum' producing images with purple or green fringes. Blair in his experiments tried to get rid also of the secondary spectrum by replacing Dollond's doublet with triplet compound lenses in which a fluid lens containing hydrochloric acid mixed with salt solutions was interspaced between a plano-convex and a converging meniscus lens of crown glass. Choosing appropriate curvatures for the spherical surfaces of the various lenses and suitable concentrations of fluids Blair was able to construct what he called 'aplanatic' compound lenses in which chromatic and spherical aberrations were reduced to a minimum.

Using his compound lenses and working both on his own and in company with his optician son Archibald, Blair was able to construct several refracting telescopes with apertures of between three and four inches. These received favourable reports from astronomers and others. Sir David Brewster writing in the *Edinburgh Encyclopedia* praised 'the ingenious labours of Dr Blair in the construction of fluid achromatic object glasses, and the high degree of perfection which he gave to the telescopes which he had constructed'. According to Brewster Blair's experiments had never received their due share of praise. However, there were many critics who pointed at the likely effects of gradual evaporation and loss of transparency of fluids and at the unavoidable corrosion of lens surfaces. Their criticisms were to some extent answered by Archibald Blair who claimed to have constructed fluid lenses which had not exhibited any deleterious effects after periods of more than twenty years.

Interesting as they were his optical experiments constituted Blair's only scientific achievement. In 1793 he moved altogether away from this field of work on his appointment as 'Emeritus First Commissioner of the Admiralty's Board for the Care and Custody of Prisoners of War'. As such he had to visit naval establishments all over the country and had to be away from Edinburgh for long periods of time. The Board on which Blair was joined by Gilbert (later Sir Gilbert) Blane was particularly concerned with the improvement of the health of sailors and Blair's contribution to the work of the Board consisted of his finding a method for the preservation of lime juice during long voyages which played an important role in the prevention of scurvy.

Blair's later semi-scientific efforts were confined to philosophical speculations which he published in 1818 as *Essays on Scientific Subjects* and in

1826 as *Scientific Aphorisms*.

From the University of Edinburgh's point of view Blair's appointment as Professor of Astronomy had been anything but happy. Blair refused to give any lectures on the grounds that he had no observatory or instruments at his disposal. Absolving himself in this way from lecturing duties Blair also kept away from other University business. Records show that after his formal induction Blair never attended any meetings of the University Senatus and indeed looked at his post as a sinecure pure and simple. He was treated with kindness, however, by some of his colleagues as is shown by the evidence which Professor Wallace gave in 1826 before the Commission on Scottish Universities when he said: 'I have no doubt that Professor Blair would have executed most faithfully the duties of his office. At the time of his appointment he was a zealous student and cultivator of astronomy and optics; but he could not carry his views into execution because Government declined to erect an observatory for the use of the University.'

Blair died after a long illness in December 1828 at Westlock in Berwickshire. His son Archibald continued his father's experiments on aplanatic telescopes until his own death eight years later. Blair's memory is kept alive in the University by a handsome bust presented by Major J. Donaldson, his wife's nephew, in 1864 which is placed amongst others in one of the bays of the University's Upper Library.

The Royal Commission on the affairs of the Scottish Universities which had started its work in 1826 did not complete its recommendations until 1831. When the Commissioners came to a discussion of the future of the Chair of Practical Astronomy following Blair's death they recommended that the vacant chair could not be filled 'until a suitable observatory, attached to the University, could be provided'. Their recommendation prevented, for the time being at least, the appointment of a remarkable astronomer who had presented himself as a candidate for the post early in 1829 and whose application had been strongly supported by some of the most eminent astronomers of the day. This candidate was Thomas Henderson who five years later was to become not only the Regius Professor of Astronomy, but also the first Astronomer Royal for Scotland.

3

THOMAS HENDERSON
FIRST ASTRONOMER ROYAL

WIDELY KNOWN at the time of his appointment for his outstanding skills in the field of astronomical computation, Thomas Henderson had never received any formal academic training in astronomy. He was born on 28th December 1798 as the youngest of a family of five. Destined like his eldest brother for the profession of the Law he had been educated first at the local grammar school and then at Dundee Academy whose Rector, a distinguished mathematician who was later to become Professor of Mathematics in the University of St Andrews, considered Henderson to have been 'one of the best scholars he ever had under his care'.

Leaving the Academy at fifteen Henderson entered a solicitor's office in Dundee where he spent the following six years classifying ancient records of the Burgh and using all his spare time to pursue his favourite hobby – astronomical calculations.

Having completed his legal apprenticeship in 1819 Henderson left Dundee for Edinburgh where his influential friend Sir James Gibson-Craig soon secured an appointment for him as advocate's clerk to the celebrated John Clerk who under the title of Lord Eldin became one of the Judges of the Supreme Court of Scotland. When Lord Eldin retired from the Bench Henderson became Private Secretary first to the Earl of Lauderdale, the leader of the Whigs in Scotland, and later to Francis Jeffrey, the Judge and literary critic who on the return of the Whigs to power in 1830 was to become Lord Advocate.

There is ample evidence that in the twelve years between 1819 and 1831 Henderson pursued his legal duties assiduously while devoting most of his leisure hours at the same time to astronomy. His record books preserved in the archives of the Royal Observatory Edinburgh contain lists made out in his neat hand of the various legal cases which were due to come up to Court interspersed with astronomical computations.

As Secretary to the Earl of Lauderdale Henderson had to make frequent visits to London where he had opportunities of being introduced to some of the leading astronomers of the time. Amongst them was Sir James South who placed his well equipped observatory at Camden Hill entirely at Henderson's disposal during his visits to London. In Edinburgh Henderson's enthusiasm for astronomy was strongly encouraged by the two Professors of Mathematics at the University, John Leslie and William Wallace. The latter in particular gave Henderson free access to the instruments of the Astronomical Institution's observatory on the Calton Hill, the 'Royal Observatory' since the summer of 1822. Using these instruments Henderson was able to gain practical experience in methods of astronomical observation. His chief interest in those years remained, however, centred on the art of astronomical computation.

Henderson wrote his first scientific paper in 1824. This was concerned with a new method for the calculation of the times of stellar occultations by the Moon and was considered sufficiently important by Thomas Young who was then Secretary to the Board of Longitude and Superintendent of the *Nautical Almanac* that he had the paper printed as an annex to each of the annual Almanacs between 1827 and 1831. An account of some of his computational methods was also published by Henderson in the *London Quarterly Journal of Science* which printed a set of papers concerned with calculations of various astronomical phenomena in the years between 1824 and 1827.

In 1827 Henderson submitted his first paper to the Royal Society for publication in its *Philosophical Transactions*. In this paper he dealt with the question of the exact difference between the longitudes of the Greenwich and Paris observatories in which Sir John Herschel had taken a great interest. There existed certain small discrepancies between the relevant observations of longitude which had worried Herschel, but being small the differences had been put down to errors of observation. In his paper Henderson could show that in fact the calculations had contained an undetected error amounting to one second of time which when allowed for raised in Herschel's words 'a result liable to much doubt to the rank of a standard scientific datum, thus conferring on a national operation all the importance it ought to possess'.

In the following years Henderson continued to exercise his remarkable talents on the solution of a wide range of computational problems including that of the determination of geographical longitude from 'Moon-culminating stars', that is stars with nearly the same right ascension and declination as the Moon at the time of observation. At the request of the Royal Astronomical Society Henderson drew up a special list of such stars which were to be used in the determination of geographical positions by the Arctic expedition of 1829 in which Sir James Ross sailing in the *Victory* was able to locate the position of the north magnetic pole.

Henderson's scientific papers, all written in his spare time while practising as a lawyer made him well known throughout the astronomical community and when he showed an inclination to exchange his legal career for that of a professional astronomer he found himself strongly supported by astronomers like Sir John Herschel who was then President of the Royal Astronomical Society.

The vacant Regius Chair in Edinburgh provided a very attractive opening to the Scot Henderson for which he presented himself as a candidate early in 1829. However, as has already been explained, the Commissioners who were responsible for the filling of the vacancy had come to the conclusion that no appointment should be made before the exact duties attached to the Chair and the question of the creation of an astronomical observatory had been considered further.

Another opening which would have been appropriate to Henderson's particular interests presented itself a few months later in the summer of 1829 as the result of the untimely death of Thomas Young, the Superintendent of the *Nautical Almanac*. The *Nautical Almanac*, originally instituted by the Astronomer Royal at Greenwich, Neville Maskelyne, in 1766, was published annually and contained tables of astronomical positions to be used by seamen as navigational aids. Young having the highest opinion of Henderson's ability had taken the unusual step only a fortnight before his death of submitting a special Memorandum to the Board of Admiralty in which he declared that he knew of 'no person more competent than Henderson' to succeed him in his post. The Admiralty however had come to the conclusion that the post of Superintendent should revert to the Astronomer Royal.

Two years later Henderson was at last successful in his efforts to enter the ranks of professional astronomers. This time the vacancy was caused by the death of the Rev. Fearon Fallows, the first Director and 'H.M. Astronomer' at the Cape of Good Hope Observatory which had been established in 1820, but which had only been completed ten years later. To Henderson the post had not the same attractions as the two earlier ones, but he accepted the appointment and sailed for South Africa in January 1832, taking up residence at the Observatory in the following April.

As things turned out, Henderson was to stay only a little over one year at the Cape, but he managed to complete in this short period of time a remarkably substantial programme of observations. These included positions of the Moon and Mars, numerous lunar occultations of stars and eclipses of Jupiter's satellites, and the paths of the comets Encke and Biela when they were visible in the southern sky. He embarked on a major programme of observing the transits of several thousand southern stars for the determination of their precise positions in the sky. His efforts were all the more remarkable as he had only one assistant, Lieutenant W. Meadows, and as his equipment left much to be desired. His instruments consisted of a Dollond 10-foot Transit Circle and a 6-foot Jones Mural Circle whose poor performance had already troubled his predecessor and whose ever necessary checks absorbed much of Henderson's time.

To summarise Henderson's work in South Africa it may be worth quoting the words of Sir David Gill, a Scot like Henderson and a later distinguished Director of the Cape Observatory: 'Henderson gave to the world a catalogue of the principal southern stars of an equal accuracy with the work of the best observatories in the northern hemisphere, and which will in all time be regarded as the true basis of the most refined sidereal astronomy of the southern hemisphere. His observations gave by far the most accurate determination of the Moon's parallax then available; they determined the longitude of the Cape with a precision which refined modern methods have barely changed.'

All these important investigations including the work for which Henderson became best known in the history of astronomy, namely his pioneer measurement of the distance of a star, were published by him only after his return to Britain from South Africa. Indifferent health and a certain dissatisfaction with his surroundings made him resign early from his office at the Cape. In May 1833 he returned to his native Scotland where he settled in Edinburgh and now free from official duties busied himself with the reduction of his Cape observations.

In the meantime discussions had taken place between members of the Government, of the University of Edinburgh and of the Astronomical Institution concerning the future of the vacant Regius Chair of Astronomy and that of the Institution's observatory. In 1834, a year after Henderson's return from the Cape, these discussions led to a formal agreement whereby the Astronomical Institution led by its President General Sir T. Makdougall Brisbane made over to the University of Edinburgh 'for their unlimited use' the Institution's 'Scientific Observatory' on condition that the Government would step in and convert this observatory into a public establishment with a Principal Observer at its head and an Assistant to support him. It was further agreed that the office of Principal Observer should be combined with that of the Regius Professor of Astronomy and that the holder of the joint post should in future have the title of 'Astronomer Royal for Scotland and Regius Professor of Astronomy in the University of Edinburgh'. While at this stage the members of the Astronomical Institution still wished to preserve their rights to the use of their 'Popular Observatory' consisting of the Gothic Tower and a small Transit-Room nearby, they agreed twelve years later to consign their remaining rights into the hands of the Government.

When the new joint Edinburgh post was announced in 1834 one of the candidates who presented himself for the post was the great man of letters Thomas Carlyle. His grounds for doing so were that some twenty years earlier he had been a student of mathematics at Edinburgh University under Professor John Leslie. However, he had left the University without ever taking a degree and when he applied for the new astronomical post, Lord Jeffrey whom he knew as Editor of the *Edinburgh Review* declined to support his candidature. Carlyle ever after thought himself ill-used and recorded his sense of injury in his later *Reminiscences*.

The post, however, was no longer to be a sinecure. When the question of making the appointment reached the Home Secretary Lord Melbourne, the Royal Astronomical Society came out in strong support of Thomas Henderson and their advice was soon accepted. The formal submission of Henderson's name to Lord Melbourne, the Secretary of State, was made by Sir Thomas Makdougall Brisbane, recently elected President of the Astronomical Institution in succession to Lord Napier and also President of the Royal Society of Edinburgh. Sir Thomas Makdougall Brisbane was himself an astronomer of high repute. He spent a number of years in Australia as Governor of New South Wales where he established at Parametta a very active private observatory. On his return to Scotland he continued his astronomical work at his home in Makerstoun, Roxburghshire, and was an important figure in the scientific life of Edinburgh. On 18th August 1834 Henderson became thus the first Astronomer Royal for Scotland and also Blair's successor as Professor of Astronomy in the University of Edinburgh. The Royal Warrant, whose wording remained unchanged for over 120 years in the appointments of six of Henderson's successors, required him 'to take upon himself the care and custody of all instruments within the Observatory of Edinburgh which belong to Us and to apply himself with diligence and zeal to making astronomical observations at the said Observatory for the extension and improvement of Astronomy, Geography and Navigation and other branches of Science connected therewith'.

Following his appointment to a post which had every possible attraction for him Henderson lost no time in starting systematic observations with the Transit Instrument and Mural Circle on the Calton Hill. It so happened that during his very first year in office Halley's comet, which returns only once every 76 years, became visible and observations of this comet were among the earliest made by Henderson on Calton Hill. He undertook a major observing programme concerned with the positions of both planets and stars, with the aim of providing data of motions in the solar system and, in the case of the stars, to compile an extensive positional catalogue for future studies of stellar movements. In his ten years in office before his untimely death he managed, with the help of only one assistant, Alexander Wallace, a graduate of Edinburgh University and the son of Professor William Wallace, to collect no fewer than 60,000 star positions. The Board of Visitors of the Observatory, which included the newly appointed Astronomer Royal at Greenwich, G. B. Airy, were lavish in their praise of his exertions and insisted on the speedy publication of his results. The first five substantial quarto volumes of the *Edinburgh Astronomical Observations* which refer to the observations made between 1835 and 1839 and which are accompanied by detailed descriptions of the exact procedures used in the observations and reductions, were published by Henderson during his lifetime; the remainder appeared after his death.

The prodigious amount of observational duties which Henderson undertook made it essential for him to live in proximity to the Observatory. For various reasons, including the expense of building on the site, it was not possible to provide a residence for him within the Observatory grounds, as was the case in most observatories. However, in 1840 an official residence was purchased for him not too far from Calton Hill at 1 Hill Crescent.

The considerable amount of work which the Edinburgh programme entailed did not induce Henderson to relinquish the reduction and publication of the observations which he had made at the Cape. Some of these have already been mentioned, but the most memorable ones, those of the binary Alpha Centauri, were by far the most important. This star, the third brightest in the sky, is a prominent object at the Cape where, being circumpolar, it can be observed at all times. The star is not only very bright, but has also the unusually large proper motion of 3.6 arc seconds per year. When Henderson became aware of this fact, unfortunately only near the end of his stay at the Cape, he came to the conclusion that Alpha Centauri might well be relatively close to us and that it would therefore be worthwhile scrutinising the observations of its position for the effect of annual parallactic motion. At the time no parallaxes had yet been measured successfully by anyone though astronomers such as Pond in Greenwich and Brinkley in Dublin had made considerable efforts trying to discover some.

Henderson's observations of the declination of Alpha Centauri though not made originally for the specific purpose of finding parallactic motion had been secured at different times of the year and were thus quite suitable for the task. When Henderson reduced these observations in Edinburgh he found indeed fairly clear evidence for the existence of a parallax amounting to about one arc second. Though his reductions were encouraging enough the whole problem of the discovery of parallax was too important for Henderson to be fully satisfied. He decided to defer publication of his findings until he had completed the reduction of the observations of the star's right ascension. Only when the results of this reduction entirely confirmed his earlier work and after much further thought did Henderson feel sufficiently confident to announce to the Royal Astronomical Society of London in January 1839 that his observations indicated a parallax for Alpha Centauri of 1.16 arc seconds with a probable error of about ten per cent.

The observation of stellar parallax, in other words the measurement of stellar distance, was the greatest advance in astronomy since the work of Tycho Brahe and Kepler two centuries earlier. Unfortunately for Henderson, his announcement had been delayed too long to allow him to claim absolute priority for this major discovery. However, the measurement of the very first parallax, that of the star 61 Cygni, which had been announced only three months earlier in October 1838, was the work of Friedrich Wilhelm Bessel of Königsberg, an astronomer for whom Henderson had

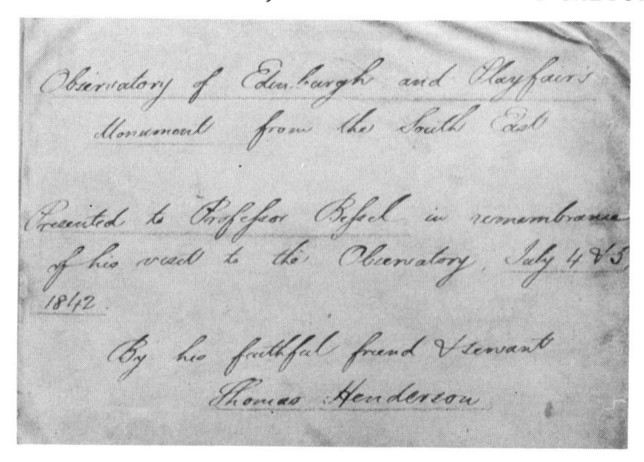

Observatory of Edinburgh and Playfair's
Monument from the South East

Presented to Professor Bessel in remembrance
of his visit to the Observatory, July 4 & 5
1842.

By his faithful friend & servant
Thomas Henderson

Thomas Henderson's handwriting on the back of a daguerrotype of the Calton Hill which he presented to F. W. Bessel (Royal Observatory Edinburgh).

always had the very greatest admiration and with whom he was to be ever after on the friendliest terms. Indeed a visit to Scotland in 1842 by Bessel in the company of the mathematician K. G. J. Jacobi, was one of the highlights in Henderson's life. The names of both Henderson and Bessel are inscribed on the ornamental stonework of the main Observatory building on Blackford Hill, as an appropriate memorial to their contribution to astronomy.

The extensive observational programme in the Observatory on which Henderson embarked when he came to Edinburgh left him little time for his duties as Professor in the University. He did not feel able to institute a formal course of lectures in astronomy, but he took over for more than a whole session lectures in mathematics and natural philosophy when his colleagues, Professors Wallace and Forbes, were unable to give them.

Henderson was never physically particularly robust and there is little doubt that his unremitting intense activity had an adverse affect on his health. A happy marriage which he contracted in 1836 with the eldest daughter of Alexander Adie, a well-known and celebrated Edinburgh optician, sustained him while it lasted, but the shock of her death in 1842, a few weeks after the birth of their only daughter, affected him all the more. Two years later his health deteriorated and on 23rd November 1844 Henderson died in his 46th year after only ten years in his Edinburgh office. Henderson's last months have been described by his successor:

> 'During the summer his health had been gradually declining and latterly from the nature of the complaint under which he laboured – disease of the heart – to ascend the Calton Hill became daily more painful and laborious. Still, such was his love of science and such his desire to discharge the duties of his office that, so long as it was possible for him at all to reach the Observatory, he continued day after day at his post and it was only when, utterly prostrated, he was at length compelled to succumb, that he ceased from those labours in the performance of which he had for so many years found his greatest pleasure.'

Thomas Henderson, who was called by his successor 'the one, sole, grand practical Astronomer that Scotland has ever produced', fulfilled indeed Lord Napier's vision of the ideal astronomer:

> 'The path of the zealous astronomer is only to be traced in the heavens; he must be abstracted from all sublunary considerations and yield himself a willing slave to the discernment of those more sublime operations of nature which transcend the thoughts or the genius of ordinary men.'

4

CHARLES PIAZZI SMYTH AND MOUNTAIN ASTRONOMY

HENDERSON'S ALL too short tenure of the Edinburgh post was succeeded by the longest and in some ways most spectacular in the history of the Observatory, that of Charles Piazzi Smyth, which lasted from 1846 to 1888. Piazzi Smyth was born at Naples on 3rd January 1819, the second son of Admiral William Henry Smyth. The Admiral had been hydrographer in the Royal Navy and while on the Mediterranean station had become a close friend of the Italian astronomer Giuseppe Piazzi of Palermo who had discovered on 1st January 1801 the first of the minor planets which he named Ceres after the tutelary deity of Sicily. Piazzi acted as godfather to the infant Charles expressing the hope that the boy might one day become an astronomer. This was very much in line with the Admiral's own ideas who after his retirement from the Navy settled in 1829 in Bedford where he built a private observatory of his own and made the observations which are collected in his classic *Cycle of Celestial Objects*, a book which created an intense popular interest in astronomy at the time.

Charles Piazzi Smyth, who always made the point of including his godfather's name in his own, grew up as a member of a very remarkable family. His elder brother was to become one of the leading British geologists while his younger brother rose to the rank of General in the Army. One of his sisters was to marry the Director of the Natural History Museum in South Kensington and another became the wife of Professor Baden-Powell of Oxford and the mother of the later Lord Baden-Powell, the hero of Mafeking.

Charles received his early education at the Bedford Grammar School where he spent six years. Instead of going up to University afterwards he left for South Africa in October 1835 where at the age of only sixteen he was to act as Assistant to Thomas (later Sir Thomas) Maclear who had succeeded Henderson as Director of the Cape Observatory. His was an unusual appointment, but his enthusiasm and his exceptional talents impressed everybody, not least Sir John Herschel, an old friend of his father's, who was at the time engaged in 'sweeping' the heavens of the southern hemisphere with his 20-foot reflector from his station at Feldhausen under the Table Mountain.

The appearance in the southern hemisphere early in 1836 of comet Halley offered the young Piazzi Smyth an opportunity to display the very considerable artistic skills for which he became well known in later life. Volume 10 of the *Memoirs of the Royal Astronomical Society* shows a number of remarkably beautiful drawings of the comet as seen in South Africa for which Piazzi Smyth was responsible.

Piazzi Smyth's life-long capacity for hard work and his strong physical constitution became apparent when he was asked to assist in the observations with the Meridian Circle of Maclear who did not scruple to work his young assistant for eighteen hours of the day. In the Observatory's geodetic work where the necessary triangulations involved lengthy and difficult observations in the wintry, snow-capped mountains to the north of the Observatory, Piazzi Smyth laboured with equal willingness and fervour.

It was to become clear in the later life of Piazzi Smyth that with his exceptional intelligence his interests could never be confined to the pursuit of pure routine work. When in March 1843 the great daylight comet of that year became visible in the southern hemisphere he made a valiant attempt to test the effect of polarisation in the comet's light – much to the disapproval of Maclear for whom proper astronomy was concerned with nothing but the measurement of celestial positions and motions and certainly not with what he termed 'physical descriptions'. However, in spite of his Director's negative attitude, Piazzi Smyth using a 3-inch portable telescope managed to obtain a fine series of observations of the comet and of the way it changed in appearance in the course of seven weeks. He made a painting of the comet's remarkably long tail which was bright enough for its reflection to be clearly visible in the sea.

While he was at the Cape Piazzi Smyth developed a special interest in the then new art of photography to which he was to make noteworthy contributions later. With the encouragement of Sir John Herschel he achieved his own production of photographs on paper and in 1844 and 1845 he made what are the oldest known photographs of scenes in South Africa including a photograph of the Cape Observatory which may indeed be the earliest photograph of any observatory anywhere in the world.

Piazzi Smyth's intellectual brilliance must have impressed itself sufficiently strongly on leading astronomers of the day such as Sir John Herschel that they were able to persuade the Court of a very traditional University like Edinburgh to appoint to an important Chair a young man of only twenty-seven who had had no formal academic training of any sort. Considering the fact that the University authorities were determined not to repeat the mistake which they had made when they appointed Dr Blair they must have been very sure of his ability when in 1846 they chose Charles Piazzi Smyth to succeed Henderson to the joint post of Professor of Practical Astronomy and Astronomer Royal for Scotland under a Royal Warrant from Queen Victoria in terms exactly similar to Henderson's.

When Piazzi Smyth arrived in Edinburgh from South Africa he found two problems confronting him. The first concerned the question of what to do with the large number of observations of star positions, some 30,000 in

Charles Piazzi Smyth. Portrait by John Faed, RSA (Royal Society of Edinburgh).

all, which had been left unreduced by his predecessor. The second was the need for the establishment of a systematic University course in Astronomy. Completing the reduction of Henderson's observations was bound to entail a massive amount of work, but Piazzi Smyth decided that it was his duty and his first priority to undertake the task. The last 5th volume of the *Edinburgh Astronomical Observations* which Henderson had published himself in 1843 contained the observations which he and his assistant, Alexander Wallace, had made in 1839. With the help of Wallace who remained chief assistant at the Observatory until 1880 Piazzi Smyth managed to reduce and publish all of Henderson's unreduced observations which appeared in five further volumes of the *Edinburgh Astronomical Observations* between 1847 and 1852. On assuming office he planned to allow himself five years in which to complete the reduction of Henderson's observations before undertaking his teaching course. In the event he achieved his target a year ahead, and began his lecturing duties in 1850.

In setting out on his own programme of observations of star positions, determining right ascensions with his transit instrument and declinations with the mural circle, Piazzi Smyth was hoping to attain a significant improvement in accuracy. He made a thorough investigation of the lack of stability of the transit instrument, something which had caused concern to Henderson because of the errors which it could produce on the measurement of star positions. He was able to trace instabilities to variations of temperature and to reduce their effect by improvements to the support of the transit instrument.

Piazzi Smyth's ultimate aim in this work was the publication of an ex-

Left: The first firing of the time gun from Edinburgh Castle in June 1861. *Right*: The firing of the one o'clock gun on the occasion of the centenary in 1961.

tensive catalogue which would contain all Edinburgh observations of star positions, both those made by Henderson and Wallace in the years between 1835 and 1845 and those made under his own direction in the years that followed. He achieved this in Volumes XIV and XV of the *Edinburgh Astronomical Observations* which appeared in 1877 and 1886. These volumes contain on 1675 pages the positions of 3890 stars as they had been determined at different times at Edinburgh and also at nine other observatories 'of acknowledged eminence'. This catalogue was to provide a kind of history of each star from which really reliable values could be deduced of the positions of the stars and of their motions. It was indeed a first attempt at the creation of a 'history of the fixed stars'.

Much of Piazzi Smyth's Edinburgh work is summarised in the Reports he presented to the members of a Board of Visitors at their annual visitations and which, written always in a very lively style, he published in the various volumes of the *Edinburgh Astronomical Observations*.

The Board of Visitors had been appointed by the Government following the handing over by the Astronomical Institution of their remaining rights of property in 1846. One of the duties which were laid onto the then fully established Royal Observatory was the inauguration of a public time service. This was accomplished after much delay with the erection in 1858 of a Time Ball on the top of the tall column of the Nelson Monument on Calton Hill and in 1861 with the mounting in Edinburgh Castle of a time gun which like the time ball was controlled by the Transit Clock of the observatory. The clock itself, 'the best astronomical clock' which could be made at the time by Messrs Dent of London, had been presented to the observatory by Sir Thomas Makdougall Brisbane. The installation of the time-gun involved a considerable amount of work. To begin with, the gun could not be mounted on Calton Hill itself on account of the delicate instruments in the Observatory. By arrangement with the Army a location for the gun was found on Edinburgh Castle and an electrical connection between the Observatory and the Castle was established. Experiments were also performed to determine the correct amount of gunpowder required to make the gun heard throughout the city without at the same time shattering nearby windows.

When everything was ready, Piazzi Smyth decided to inaugurate the new time service formally on 5th June 1861 in the presence of his Board of Visitors and a large gathering of distinguished Edinburgh citizens. Amongst guests from outside was Piazzi Smyth's astronomer friend from Dublin, Sir William Rowan Hamilton, the Astronomer Royal for Ireland. At precisely one o'clock the time-ball dropped correctly from the Nelson Monument, but there was no sound from the gun, much to the embarrassment of Piazzi Smyth who was not accustomed to failures in his technical endeavours. However, the fault was rectified within a week, and thereafter the time-gun became well known for its dependability. Later the Calton Hill Observatory also controlled time-guns at Glasgow and Newcastle-upon-Tyne. The time-gun has remained a well-known feature in the life of Edinburgh where it is still fired daily from the Castle at one o'clock.

Piazzi's Smyth's interests were always wide ranging. Among his early ones on Calton Hill was a discussion of the temperature readings of a number of rock thermometers which had been inserted into the ground of the observatory at various depths down to 24 feet for an investigation of the effects on earth temperatures of seasonal and secular variations of solar radiation. Piazzi Smyth studied earth temperatures for many years in an attempt to correlate them with the eleven-year sunspot period. The observations which had started in Henderson's time in 1836 came to an end only in September 1876 when according to Piazzi Smyth's Report for that year a mad sailor from a Portuguese ship entered the observatory and broke all the thermometers!

In those early years Piazzi Smyth also found time to invent new types of astronomical instruments. Among them was a 'free-revolver stand', built on the principle of the gyroscope for the steady mounting of instruments at sea. Another was the 'Edinburgh Universal Instrument' which could be used as a sextant at sea and as a theodolite or transit on land. It was recommended by Piazzi Smyth for the work of students in his Class of Practical Astronomy. At the Paris Universal Exposition of 1855 he exhibited no less than twenty 'New or Improved Instruments for Navigation and Astronomy', designed by himself, the prototypes constructed locally mainly by the optician John Adie of the same family to which Thomas Henderson's wife belonged. A descriptive Catalogue of Piazzi Smyth's exhibits was published by the University of Edinburgh.

Volume XI of the *Edinburgh Astronomical Observations* contains, in addition to a record of meridian observations made between 1849 and 1854, a detailed account of an expedition to the west coast of Norway which Piazzi Smyth undertook in company with Dr T. R. Robinson, the Director of the Armagh Observatory in Ireland, in order to observe the total solar eclipse of July 1851. The weather turned out to be cloudy at the critical time, but it was typical of Piazzi Smyth that the occasion was not wasted; he was able to make a number of sketches of the effects on sky and countryside of the eclipse before, at and after totality. Two of these very remarkable coloured sketches were included in this particular Volume XI and were later reproduced by Sir Robert Ball, the most popular astronomical author of the day, in his *Story of the Sun* of 1893. The same volume of the *Edinburgh Observations* contains another example of Piazzi Smyth's artistic flair in the form of drawings of the Lunar Region near Mare Crisium which he undertook at the request of the British Association for the Advancement of Science.

In his Annual Report for 1852 we come across for the first time Piazzi Smyth's idea of 'Mountain Astronomy' to improve the effectiveness and precision of astronomical observations by moving telescopes to high altitudes, to Newton's 'most serene and quiet air, such as may perhaps be found on the tops of the highest mountains above the grosser clouds'. It was in this Report that he mentioned in particular the Peak of Tenerife as a place which he judged would be very suitable for the establishment of an Edinburgh observing station for the summer months. He suggested the Peak because of its high elevation of some 12,000 feet which could yet be ascended easily and also because of its low geographical latitude of 28 degrees which would allow observations of stars to be made in the middle of the summer when twilight makes such observations almost impossible in the high latitude of Edinburgh.

What in 1852 had been a happy idea in Piazzi Smyth's mind became an actual programme of work four years later when he received the support of George Biddel Airy, the Astronomer Royal in Greenwich, who pressed upon Government the scientific importance of Piazzi Smyth's ideas and suggested that the necessary tests should be entrusted to the Royal Observatory in Edinburgh. As a result, Piazzi Smyth was officially asked by the Admiralty to lead a scientific mission to Tenerife to ascertain 'how much astronomical observation can be benefitted by raising telescopes high into the air, thus eliminating the lower third or fourth part of the atmosphere'. Following suggestions by the Royal Society, the Royal Astronomical Society, Sir John Herschel and other scientists the expedition was also to concern itself with certain physical, meteorological, geological and botanical observations while it was on Tenerife.

Piazzi Smyth's proposal was formally sanctioned by the Admiralty on 1st May 1856 and once this had been done no time was lost by him in collecting a variety of instruments, meteorological as well as astronomical. There were two telescopes, a small 3-inch refractor belonging to the Observatory and a 7-inch Cook equatorial which was loaned to Piazzi Smyth by Mr H. L. Pattinson of Newcastle upon Tyne. Among other instruments was a 'thermomultiplier' by Gassiot, probably similar to the thermopile which Melloni had developed in 1846, which Piazzi Smyth intended to use for the measurement of the thermal radiation of the Moon.

In June everything was ready for the expedition to sail to Tenerife in the yacht *Titania* of 140 tons which with its crew of sixteen men had been placed at the disposal of Piazzi Smyth by one of his admirers, Robert Stephenson, a Member of Parliament. This time Piazzi Smyth was accompanied by his young wife whom he had married on the previous Christmas Eve. His wife, Jessica Duncan, was in fact to become his most enthusiastic and faithful assistant in all his endeavours during nearly forty years of their married life and accompanied him on all his subsequent travels.

The expedition arrived in Santa Cruz on Tenerife on 8th July. A week later, serenaded by a military band and with a cavalcade of twenty mules they marched from Orotava (now Puerto de la Cruz) to the 8900-feet summit of Guajara where the party camped for five weeks before moving on to Alta Vista at 10,700 feet on the slopes of the Peak of Teide, the highest point accessible to mules. They descended to sea level in September, re-embarked in the yacht and returned to Southampton on 14th October.

This pioneer 'seeing expedition' during which Piazzi Smyth spent 65 days at the two high-altitude stations yielded some very remarkable results. Using his 3-inch telescope at the Guajara station he found the clarity of sky to be such that he could readily see stars of 14th magnitude, four magnitudes fainter than the faintest ones he had been able to observe with the same telescope in Edinburgh. He noticed at the same time the outstanding definition and perfect steadiness of the stellar images in his telescope. The magnificent seeing conditions which he found both at Guajara and Alta Vista made him re-observe for separation and position angle the components of some fifty double stars. With his 7-inch telescope at Alta Vista he was able to see the faint companion of the star Antares twenty minutes before sunset and to separate readily double stars as close as one arc second.

Top: The station at Guajara, Tenerife. The stereophotograph was taken by Piazzi Smyth and shows his wife and assistant Jessica. *Bottom:* The Alta Vista station in Tenerife, another of Piazzi Smyth's stereophotographs of his site-testing equipment.

At Alta Vista he also managed to obtain some remarkable observations of the planet Jupiter which presented 'a sight very different from anything that European astronomers have had for many years past'. Three beautiful drawings of Jupiter – 'they might almost be called revelations' to use his own expression – can be found in Volume XII of the *Edinburgh Astronomical Observations*.

Piazzi Smyth on Tenerife was the first to obtain a quantitative estimate of the thermal radiation of the Moon. Using his Gassiot thermomultiplier at the Guajara station he compared the radiation received from the Full Moon with that radiated by a candle. In order to allow in his measurements for the effect of radiation coming from the sky he pointed his thermopile alternately

Lines in Red ends of sun-spectrums, direct and reflected.

No. for Refs.	Day	Hour	Direct or Reflected	Alt. of Sun	Height of Station	Estd Value
					feet	
1	Sept. 23	0h PM	D.	61°	50	10
2	Augt. 7	0 PM	R.	78	8903	4
3	Augt. 8	0.30 PM	R.	77	8903	7
4	Augt. 8	8.30 AM	R.	20	8903	5
5	Augt. 9	5.30 AM	D.	5	8903	4
6	Augt. 9	PM	D.	23	8903	4
7	Augt. 9	PM	D.	1.0	8903	4
8	Augt. 8	PM	D.	-0.1	8903	2
9	Augt. 12	AM	D.	-0.5	8903	5
10	Augt. 9	PM	D.	-1.1	8903	4

Spectra of the Sun at various altitudes from 78° to 1° below the horizon recorded by Piazzi Smyth in Tenerife, showing the absorption bands due to the Earth's atmosphere.

at the Moon and at the sky 20 degrees to the east and west of it, becoming thereby the first observer to use the method known to modern astronomers as 'sky-chopping'.

Amongst other astronomical observations on Tenerife was a study of the spectrum of the Sun and the effect on the solar spectrum of the terrestrial atmosphere. Observing the Sun at different altitudes ranging from 78 degrees to minus 1 degree at sunset Piazzi Smyth shows clearly in his drawings how the Sun's spectrum changes on account of the presence of absorption lines produced by the Earth's atmosphere which are strengthened when the Sun approaches the horizon. In his solar observations at very low altitude Piazzi Smyth was able to extend the study of the infrared end of the solar spectrum to a wavelength of 7800 Å.

Piazzi Smyth's astronomical work was accompanied by systematic meteorological observations at the two mountain stations and at sea level where they were made simultaneously by the captain of the yacht *Titania* which was lying all the time in Santa Cruz roads. Of particular interest is his examination of the origin of the peculiar dust haze which he first noticed at Guajara and which made him move his observing station to the higher ground of Alta Vista.

Much new information was gathered also by Piazzi Smyth on the geology of the great crater of the Peak and of the various layers of its lava streams. He explored in particular the famous subterranean ice cavern which one finds at an altitude of 11,000 feet among the broken streams of black lava. Last but not least, the expedition busied itself with a study of the botany of Tenerife, with the changes of its vegetation with altitude and with the nature of the Dragon Tree, the plant most typical of the island.

The final results of this remarkable scientific expedition were published by Piazzi Smyth in the *Philosophical Transactions of the Royal Society* for

1858 and in greater detail in a special Report addressed to the Lord Commissioners of the Admiralty which was printed in 1858/59. Further details and the philosophy behind Piazzi Smyth's 'Teneriffe Experiment of 1856' are contained in Volume XII of the *Edinburgh Astronomical Observations*.

Considering the wealth of new observational material which Piazzi Smyth was able to bring home his enterprise must be looked at as one of the most successful – and with a grant of only £500 least expensive – scientific expeditions ever undertaken. Its principal object being to test the truth of Newton's opinion as to the likely advantages of mountain sites for astronomical observatories his journey to Tenerife goes down in astronomical history as the first ever scientifically conducted astronomical site-testing expedition.

Piazzi Smyth hoped that his experiences in Tenerife would lead to the foundation of an outstation 'connecting a peripatetic with the fixed observatory of Edinburgh' and allowing the observatory 'to occupy the useful and honourable position in the scientific world which is proper to the Royal Observatory of the ancient kingdom of Scotland'. These hopes were to remain unfulfilled in his time and indeed for more than a century. The Report seems to have been quickly forgotten; perhaps the scheme was too bold and unconventional to recommend itself to the dispensers of public funds and to the scientific establishment. However, the idea was destined to be revived with history almost repeating itself in a most satisfactory way. In 1967 the case for the creation of a major British national observatory in a good climate not too far from home was put forward by the present author on behalf of the Royal Observatory Edinburgh and submitted to the newly established Science Research Council where it was accepted two years later. The responsibility for finding the most suitable site for the new observatory was placed on the Royal Observatory Edinburgh whose staff began their search in 1970. After some initial studies of sites in the Mediterranean area the site-testing was concentrated on four high altitude sites, two of them in the Canary Islands. The site finally chosen is on the island of La Palma, a sister island of Tenerife where conditions of sky transparency and steadiness were found to be outstanding. The new observatory which is now being built will therefore be not far from Piazzi Smyth's original choice of Tenerife where the remains of his observing stations at Guajara and Alta Vista are still in existence.

Apart from the scientific accounts of his expedition which were mentioned earlier Piazzi Smyth also wrote a most delightful popular book on his and his wife's experiences which he called *Teneriffe, an Astronomer's Experiment: or, Specialties of a Residence above the Clouds*. This charming book which he published in 1858 and dedicated to the First Lord of the Admiralty is illustrated by a large number of his own very remarkable photo stereographs.

In the summer of 1859 Piazzi Smyth accompanied by his wife went on a voyage to Russia to visit the great Imperial Observatory at Pulkowa near St Petersburg. They sailed from the Port of Leith by way of the Baltic to St Petersburg and travelled from there to Moscow and Novgorod. The Smyths were enthusiastic travellers. Throughout the journey Piazzi Smyth recorded everything of interest in his Diary which he illustrated with numerous delightful sketches of the foreign scenes. On his return to Edinburgh he published in 1862 an account of his experiences in the two volumes of his book *Three Cities in Russia* which he illustrated with a number of his sketches. The book gives a most interesting account of the Russian scene in the middle of the 19th century, of the lives, views and traditions of the Russians with whom the Piazzi Smyths came in contact.

Of particular interest in this book to the astronomical reader is the description of the observatory at Pulkowa which had been opened only twenty years previously under the direction of the great Friedrich Georg Wilhelm Struve, the founder of the famous Struve dynasty of astronomers whose members occupied leading posts in many European observatories in the 19th century. The new Director was Friedrich's son Otto who received the visitors with the warmest hospitality and exhibited the greatest possible interest in Piazzi Smyth's ideas on 'Mountain Astronomy' and his work on Tenerife.

The interesting journey to Russia had a sad sequel. In the following October the ship *Edinburgh* in which the Piazzi Smyths had crossed the North Sea went down with all hands in a fierce storm one day out of Leith. Piazzi Smyth and his wife had seen the ship off, having handed over boxes which contained prints of his Russian stereophotographs. Piazzi Smyth had spent six weeks preparing these for dispatch to his Russian friends. The tragedy was all the more distressing for the Smyths as they had become personal friends of the ship's captain, Captain Steel, and his wife. Later Piazzi Smyth investigated from a collection of meteorological records the exact course of the storm and published his results with typical thoroughness in an illustrated report called 'A Hyperborean Storm' in the *Edinburgh Astronomical Observations*.

Disappointed by the poor response at home to his proposals for a mountain observatory, Piazzi Smyth was greatly encouraged by the very friendly reception which his work received in France and Russia, where in 1880 it led to the foundation of a French mountain observatory on the Pic du Midi in the Pyrenees at an altitude of 9400 feet, and in 1892 to a Russian mountain observatory at Abastumani in Georgia at an altitude of 4600 feet, both flourishing institutions until the present time.

5

CHARLES PIAZZI SMYTH
THE GREAT PYRAMID AND SOLAR SPECTROSCOPY

IN THE 1860s Piazzi Smyth's mind moved into an entirely new direction which after initial successes and much public interest was to distract him, at least for a while, from solid scientific work into strange mystical speculation, much to the detriment of his academic reputation. He became increasingly engrossed with the exploration of the Great Pyramid of Cheops at Giza near Cairo which he began to see as a unique metrological monument containing in its structure standards of weights and measures for all time.

The immediate cause for Piazzi Smyth's new interest was some correspondence with John Taylor, the editor of the London *Observer* who in his later years had been attracted by accounts which Colonel Howard Vyse had brought back from his exploration of the Great Pyramid and published in 1837. Taylor wrote two books on the subject, *The Great Pyramid: why it was built and who built it* and *The Battle of the Standards* both of which he sent to Piazzi Smyth for comment. Using Howard Vyse's measures Taylor had come to the conclusion that the circumference of the base of the Pyramid stood in the same relation to its height as the circumference of a circle to its radius, in other words, that the dimensions of the Pyramid implied that its builders were familiar with the value of the number Pi and in constructing the Pyramid wished to demonstrate that knowledge. Taylor also accepted the belief of E. F. Jomard, one of the French savants who accompanied Napoleon on his expedition to the Pyramids in 1798, that the builders of the Pyramid had the necessary astronomical knowledge to enable them to measure the length of a geographical degree and thereby the circumference of the Earth. This assumption led Taylor to propose that the builders of the Pyramid chose the ratio of its height to the circumference of its base in order to demonstrate their knowledge of the radius of the Earth. As he put it: 'It was to make a record of the measure of the Earth that the Pyramid was built'.

Taylor furthermore suggested that in their construction of the Pyramid the builders had measured lengths in units of a cubit which was nearly exactly 25 British inches long, or in other words that the British inch is a measure of very ancient origin.

Piazzi Smyth submitting his Annual Report for 1864 to his Board of Visitors explained how he was fascinated by Taylor's ideas and how the further he proceeded in his enquiry into Taylor's theory 'the more some of Mr Taylor's leading views seemed to commend themselves'. Later in the same Report he states 'a high probability was impressed upon my mind, that the Great Pyramid besides its tombic use, might have been originally invented and designed to be appropriate for no less than a primitive Metrological Monument'.

Piazzi Smyth wrote this at a time when the question of units of measurement was the subject of heated debate in Britain where efforts were made by one party to replace the traditional Imperial system of weights and measures by one based on metric units. In the 'battle between the inch and the centimetre' Piazzi Smyth was very much on the traditional side like his friend and mentor Sir John Herschel who went so far as to resign his membership of the Standards Commission because it favoured the introduction of the metric system. According to Herschel the French metre was an entirely unsuitable standard of length because it was defined in terms of the length of a curved meridian of the Earth. His suggestion was that the polar axis of the Earth provides a much more natural and reliable basis for a standard of length. According to the figures which Herschel had at his disposal at the time the length of the polar axis of the Earth amounted to

Charles Piazzi Smyth: a photograph taken in his Pyramid years.

500,500,000 British inches or to exactly 500 million inches if the British inch were lengthened by a thousandth part, the thickness of a human hair. If such were done the inch, according to Herschel, would become 'a truly scientific, Earth commensurable unit of length' and 25 such inches would make a very useful cubit.

Taylor's suggestion that the inch, 'this natural standard of length', had already been used as a measure by the builders of the Pyramid appealed strongly to Piazzi Smyth's mind. However, since to him a choice between Imperial and metric units was a matter of the greatest national importance he decided to go out to Egypt himself for a new exploration of the Pyramid. His aim was to re-measure with the greatest possible precision the Pyramid's external size and the dimensions of its interior passages and chambers. He was also anxious to test the evidence for an earlier suggestion of Sir John Herschel's that it might be possible to ascertain the time of the building of the Pyramid if one could assume that it had been orientated astronomically. Apart from the alignment of the four sides of the Pyramid to the cardinal directions there is a sloping entrance passage in its northern face which might have been arranged, so Herschel suggested, in such a way that it pointed to the position of the celestial pole or to a sufficiently bright star near that pole at the time when the Pyramid was built.

On account of a slow conical motion of the Earth's axis the celestial pole moves amongst the stars, and the star nearest to it at the present time, the Pole Star Alpha Ursae Minoris, was at a considerable distance from the celestial pole in ancient times. However, as could easily be shown, another fairly bright star, Alpha Draconis, was close to the celestial pole at around

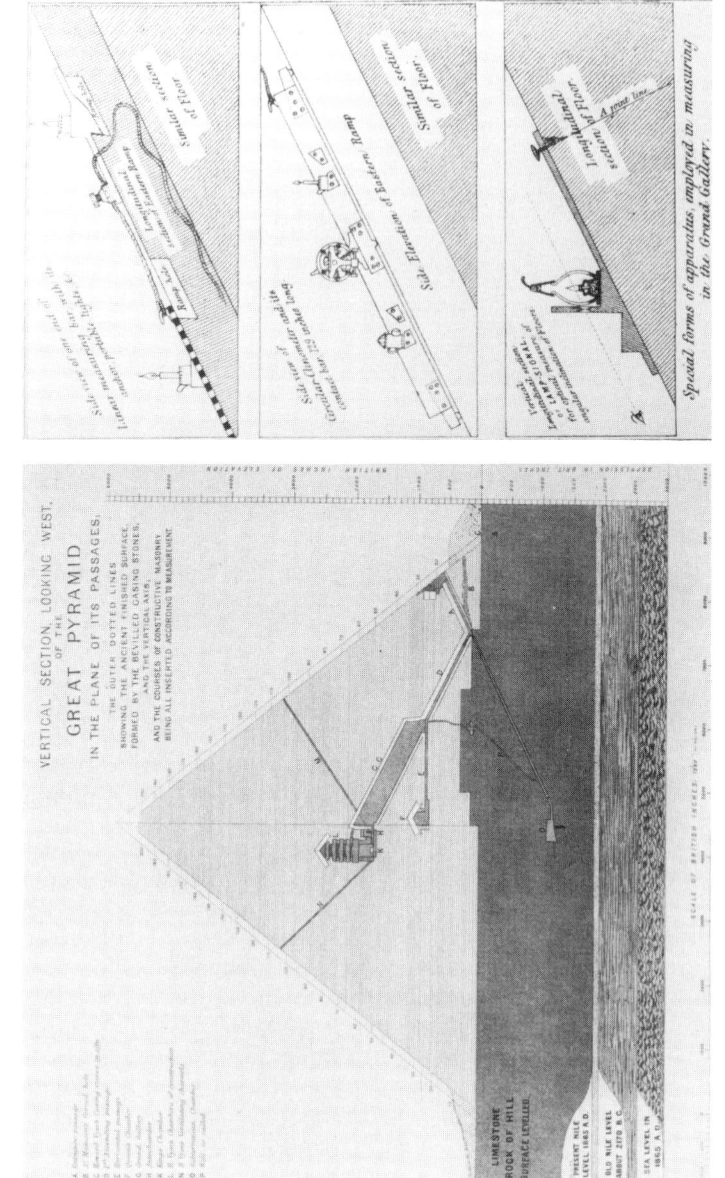

Left: A drawing of the interior of the Great Pyramid of Giza. One of the many illustrations in Piazzi Smyth's accounts of his Pyramid work. *Right*: Diagram of the apparatus used by Piazzi Smyth to measure the inclination of the interior passage of the Pyramid.

The miniature camera with its plateholder constructed by Piazzi Smyth and used for photographing the interior of the Great Pyramid (Royal Observatory Edinburgh).

2800 BC and this was the time which Herschel suggested as the likely age of the Pyramid. The Egyptologists of the time, however, thought that the Pyramid was at least one or two thousand years older, and Piazzi Smyth hoped to be able to re-examine the whole question in the course of his expedition to the Pyramid.

The expedition sailed for Egypt in November 1864. By that time Piazzi Smyth was already strongly imbued with John Taylor's opinions including the view that the great Pyramid could only be the work of chosen men who had built it under divine guidance. Piazzi Smyth's book *Our Inheritance in the Great Pyramid* of which the first edition appeared two months before he left for Egypt – and of which three more editions were to follow – shows very clearly the direction of his thinking at the time.

Piazzi Smyth and his wife arrived in Cairo with twenty-seven boxes filled with measuring equipment and general stores, 'a most extremely inconvenient quantity of packages' as the mate of the ship they were travelling on had told them. There was a variety of measuring rods of different lengths including a metal bar with built-in thermometers at each end to allow for changes in its length with changes in temperature. All measuring rods including a standard made of a piece of ancient basalt were to be ultimately compared – after Piazzi Smyth's return to Edinburgh – with a standard yard measure which had been constructed by Captain Henry Kater, well known for his accurate determination of the length of the seconds pendulum.

The inclinations of the various descending and ascending passages in the interior of the Pyramid were to be measured with a specially constructed clinometer and with an altitude-azimuth instrument which had once belonged to Professor John Playfair.

There were also meteorological instruments to allow Piazzi Smyth to set up a meteorological station near the Pyramid and to test temperatures in various parts of its interior.

Piazzi Smyth was already a skilled photographer at the time of his work on Tenerife. It was natural therefore that his equipment for the Egyptian expedition should include photographic cameras and dry and wet plates for the photography of the exterior of the Great Pyramid and its neighbours and the interior of the Pyramid's chambers and galleries. He took some 160 photographs, one half of them on dry plates about three inches square, the other half on wet plates of the unusually small size of one inch square for which he used a Dallmeyer lens of 1.8 inches focal length. The plates for this first 'miniature camera' were three inches long and one inch wide, similar in size to standard slides for microscopes.

Piazzi Smyth was not only the pioneer of miniature photography, he was also one of the very first to use a flash in the form of mixtures of magnesium and gunpowder which he made up when he wanted to photograph the dark interiors of the Pyramid.

Keen on stereo photography, Piazzi Smyth used a pair of cameras which he mounted sufficiently far apart to produce the desired three-dimensional effect. Though many of Piazzi Smyth's remarkable photographs have become lost in the course of time, some are still in existence and in the possession of the Royal Society of Edinburgh. Piazzi Smyth's famous miniature camera is kept in the archives of the Royal Observatory Edinburgh.

Having received a promise of some general assistance from the Viceroy Ismail Pasha – the same who was later to commission Verdi's opera *Aida* – Piazzi Smyth and his wife set out for the Pyramids in early January 1865 accompanied by a substantial baggage train. As their residence and general headquarters at the Pyramids they chose one of the ancient tombs to the east of the Great Pyramid which had been used earlier by Colonel Howard Vyse. For the next four months this tomb was to become their living room, laboratory, workshop and store.

With the assistance of his wife and a general factotum, Ali Gabri, who had already served Howard Vyse thirty years before, Piazzi Smyth started his investigation late in January when the inner chambers of the Pyramid had been cleared of mountains of rubble. In measuring the dimensions and inclinations of the various interior passages and of the chambers of the King and Queen he paid particular attention to the 'coffer' or 'sarcophagus' in the King's chamber whose cubic content he thought had served as unit of capacity. All his findings were published by him after his return to Edinburgh in a substantial detailed Report in Volume XIII of the *Edinburgh Astronomical Observations* which he illustrated with a set of 33 remarkable 'Great Pyramid Plates'. The same Report for which the Royal Society of Edinburgh awarded him its Keith Prize in 1867 contains Piazzi Smyth's re-examination of the evidence for Herschel's dating of the Pyramid with the help of the star Alpha Draconis. He came to the conclusion that the actual building of the Pyramid was nearer 2200 BC when Alpha Draconis was in the right position below the celestial pole and the Pleiades were on the meridian above the pole. Though modern investigations have thrown grave doubts on the assumption that the entrance passage to the Pyramid was ever used for astronomical observations it is interesting to note that both Herschel's and Piazzi Smyth's datings are not inconsistent with modern archaeological evidence.

Summing up his measures of the exterior of the Pyramid Piazzi Smyth came to the conclusion that its builders had indeed used as unit of length a sacred cubit of 25.025 British inches, which according to Herschel was a useful measure since it was related to the length of the polar axis of the Earth. According to Piazzi Smyth it was even more than that: 365¼ such cubits gave the length of the base-side of the Pyramid or in other words, there were as many sacred cubits in one side of the Pyramid's base as there are days in a year!

These and other more and more surprising findings seemed to confirm and reinforce John Taylor's ideas about the profound mathematical and scientific knowledge of the builders of the Great Pyramid which 'revealed a most surprisingly accurate knowledge of high astronomical and geographical physics nearly 1500 years earlier than the extremely infantine beginning of such things among the ancient Greeks'.

Piazzi Smyth published his conclusions in great detail in the three volumes of his *Life and Work at the Great Pyramid* which appeared in 1867 and also in his book *On the Antiquity of Intellectual Man, from a Practical and Astronomical Point of View* which was published a year later. He dedicated his *Life and Work* to the memory of John Greaves who in 1638 had made an early exploration of the Pyramid and who had been rewarded for his work by his appointment as Savilian Professor of Astronomy at Oxford; and to Colonel Howard Vyse who in 1837 had been his immediate predecessor in Egypt. A particular dedication was reserved for 'Napoleon Bonaparte, Republican General in 1798, who earnestly sought to moderate the rigours of war upon the ancient land of Egypt, by causing his army to become the most efficient means for introducing there the elevating influences of science'.

The first volume of his *Life and Work* gives a fascinating account of the daily lives of Piazzi Smyth and his wife during their four months in Egypt. It is written in the entertaining style of his earlier books on their travels to Tenerife and Russia. The second volume contains the details of his various measurements and the third a discussion of his results and his attempt to interpret his data in terms of a theory of the Pyramid as a great metrological monument built with divine inspiration for the instruction and guidance of all mankind. The work and the following *Antiquity of Intellectual Man* was bound to rouse considerable popular interest, but also to stir up criticism from more than one side. Strong attacks on Piazzi Smyth's mystical ideas started at meetings of the Royal Society of Edinburgh, continued with controversies with the Ordnance Survey over their measures of the Pyramid, and led in 1874 to the unheard-of resignation of Piazzi Smyth from the Royal Society of London of which he had been elected Fellow in 1850. It was extraordinarily unfortunate that for many years Piazzi Smyth's remarkable earlier reputation as a scientist was seriously affected by his efforts in 'Pyramidology'.

It is curious that the earliest refutation of Piazzi Smyth's ideas was to come from the work of the son of William Petrie who had been one of Piazzi

Smyth's greatest admirers. William Flinders Petrie who later was to become Professor of Egyptology at University College London and the leading Egyptologist of his day, went to Egypt in 1880 to start an entirely new precise triangulation of the Great Pyramid and its neighbours. His measurements published three years later in *The Pyramids and Temples of Gizeh* showed clearly that the builders of the Great Pyramid had used as measure of length not Piazzi Smyth's cubit, but the 'royal cubit' of 20.63 inches which gave a base-line for the Pyramid of 440 and a height of 280 cubits. Though accepting the fact that the ratio of the Pyramid's circumference to twice its height is equal to the value of the number Pi, the new measures did away entirely with Piazzi Smyth's suggestion that the length of the circumference is in some way related to the length of a year. However, in spite of all the criticisms levelled against him Piazzi Smyth never abandoned his belief in a divinely inspired 'design of that most primeval and most purely scientific building of all the Earth, the Great Pyramid of Egypt'.

Though in the course of the 1860s Piazzi Smyth got increasingly carried away by his interest in the Pyramid, the Annual Reports to his Board of Visitors show clearly that he did not neglect his responsibility for the ordinary work of the observatory. Observations of star positions were carried on with the Transit Instrument and Mural Circle though the performance of the latter was found to become increasingly unsatisfactory. The public time-service with time ball and time-gun was maintained and so was the collection of meteorological records which the observatory received from 55 stations in Scotland. Volume XIII of the *Edinburgh Observations* contains a detailed discussion of the nature of the hurricane of October 1860. Piazzi Smyth was always interested in atmospheric phenomena and he was amongst the first observers to report a sky glow during a spectacular display of the Leonid meteors in November 1866 when 'there seemed to be a glow of infinite numbers of distant, and not individually visible meteors'.

Among Piazzi Smyth's many travels was a journey through Germany in 1869 which included a stay of several days in Munich spent mostly in the company of another Astronomer Royal – Professor Johann von Lamont, Astronomer Royal of Bavaria. A native of Braemar, Lamont had been brought up from the age of 13 by Scottish Benedictines at their Bavarian monastery at Regensburg and had become in 1835 when he was thirty Director of the observatory at Bogenhausen near Munich where he attained great distinction through his work in geophysics and particularly in terrestrial magnetism. Piazzi Smyth was fascinated by the number and quality of the scientific instruments which filled every available corner of the Bogenhausen Observatory. Piazzi Smyth's work on the variations of Earth temperatures made him greatly interested in Lamont's discovery – from magnetic observations made at Göttingen and Munich between 1835 and 1850 – of a distinct 10 or 11 year periodicity in the strength of terrestrial magnetism. Piazzi Smyth was greatly impressed to find such a distinguished Scot working so far away from home. And when Lord Lindsay visited Lamont a few years later he first thought of him as a Frenchman, addressing him in French until Lamont responded 'in broad Aberdeen Scotch'.

In the 1870s Piazzi Smyth started to devote much of his time and energy to the field of spectroscopy, to observations of the spectrum of the Sun and of the daylight sky which he studied for the presence of 'rain bands' as indicators of meteorological conditions. These are absorption bands in the yellow part of the spectrum caused by atmospheric water vapour. He also obtained spectra of the Aurora, of the Zodiacal light and of various gases in the laboratory. Volumes XIII and XIV of the *Edinburgh Astronomical Observations* published between 1871 and 1877 contain many results of Piazzi Smyth's solar work starting with low-dispersion spectra observed in Edinburgh and going on to high-dispersion prism and grating spectra which he studied in 1877 at the Royal Observatory in Lisbon and in 1881 near Funchal on the island of Madeira. His Lisbon investigations were also published by him in the 29th Volume of the *Transactions of the Royal Society of Edinburgh* which awarded him its Makdougall-Brisbane Prize for this work. His solar observations on Madeira appeared in 1882 in a special book called *Madeira Spectroscopic*. His most substantial study of the solar spectrum was made in the summer months of 1884 with a high-dispersion spectroscope incorporating one of the new large 'magnificent' diffraction gratings which Piazzi Smyth had acquired directly from its maker, Professor H. A. Rowland of the Johns Hopkins University, Baltimore. To avoid the smoke-laden atmosphere of Edinburgh Piazzi Smyth went south in quest of a suitable site from which to make his observations, and settled on a country spot near Winchester where he rented a house and installed his apparatus in a south-facing room on the upper floor. In this work he tried also with no definite success to trace possible effects on the solar spectrum of the peculiar atmospheric conditions which followed the volcanic eruption on Krakatoa in 1883. His results were published by the Royal Society of Edinburgh in Volume 32 of their *Transactions*, which contains on 60 coloured plates a magnificent map of the visual solar spectrum from violet to red.

In the same volume there are the last of several papers which Piazzi Smyth wrote on his experiments in laboratory spectroscopy where he was particularly interested in the CH and CO molecules and where, in association with his friend Professor A. S. Herschel, he discovered the structure of the bands of carbon monoxide. All his laboratory work was carried out in his official residence at 15 Royal Terrace at the bottom of the Calton Hill where he had moved in 1870 from his former residence at 1 Hill Crescent. It was at the house in Royal Terrace that he demonstrated his spectroscopic work to the many visitors to the observatory. Sir Robert Ball, Lowndean Professor at Cambridge and one of Piazzi Smyth's numerous friends – who had been Astronomer Royal for Ireland and whom Piazzi Smyth always addressed in correspondence as 'Dear Brother Astronomer Royal' – described in his *Reminiscences* a humorous episode during his visit. Piazzi Smyth had begun to go deaf in middle age and was completely deaf by the age of sixty. He used to carry a slate about him on which his interlocutors could write, but according to Ball 'his flow of conversation was so inexhaustible that a visitor found few opportunities for making use of the slate'. When Piazzi Smyth showed Ball his latest spectroscopic equipment including a powerful coil for an electric spark, the coil made such a deafening noise that Ball did not hear a single syllable of the lengthy explanations which Piazzi Smyth, perfectly unconscious of the noise, gave him of his experiments.

The open views of the northern sky which he had from his house in Royal Terrace gave Piazzi Smyth a great interest in Aurorae which were particularly frequent in the early 1870s due to high sunspot activity. Coloured plates in Volume XIV of the *Edinburgh Astronomical Observations* show paintings which he made of several remarkable Aurorae whose low-dispersion spectra are displayed on other plates. The same volume shows some beautiful water-colours of the Zodiacal Light as he saw it at sea and in Palermo. As in the case of the Aurorae these paintings are again accompanied by plates of low-dispersion spectra of the Zodiacal Light.

Many of Piazzi Smyth's researches were only made possible by his paying the expenses out of his private means. He complained on numerous occasions about the lack of Government funds for the running of the observatory, for the maintenance of instruments and for the necessary replacement of out-of-date equipment. Last not least, there was the inadequacy of the salaries of his own and of his one and later two assistants. Apart from his official residence he had 'as a special salary of the Astronomer Royal for Scotland' a sum of £100 p.a. which he received in addition to £300 p.a. which came to him from the University of Edinburgh. In his efforts to secure a larger Government grant for the observatory he was strongly supported by his Board of Visitors and by both the Royal Society of Edinburgh and the Royal Astronomical Society in London. However, all his applications were without success, probably as the result of adverse criticisms of the observatory's usefulness which were contained in an official Report by the Hydrographer to the Admiralty in 1872.

It was obvious that the role of the Royal Observatory was coming under scrutiny when in 1874 it was formally inspected 'without any preliminary notice' by the Home Secretary and when in 1876 a Government Commission of Inquiry under the chairmanship of Lord Lindsay was appointed by the Home Secretary 'to examine into, and report upon, everything in connection with the Royal Observatory, Edinburgh'.

The situation was not made any easier when in 1875 Piazzi Smyth started an open controversy with the University of Edinburgh over his position and duties as Professor. When he was first appointed to the Edinburgh post in 1846 Piazzi Smyth had been full of enthusiasm and keen on starting a formal course on Practical Astronomy after he had dealt with the reduction of his predecessor's observations. The course proposed by him was discussed and agreed in April 1850 by a special committee which included the Principal of the University John Lee and the Professors of Mathematics and Natural Philosophy, Philip Kelland and James David Forbes. Though the new course was to concentrate on 'practical' astronomy leaving the teaching of theoretical fields such as celestial mechanics or spherical astronomy to others, Piazzi Smyth was well content and started lecturing, the first Edinburgh astronomer to do so, in November 1850. His lecture notes, which still exist, make very interesting reading. The course, which included weekly practical exercises at the Observatory, was well prepared and in parts at least very attractive. However, he did not get much co-operation from the Natural Philosophy Department and he also found the intellectual level of his students much lower than he had expected. Their numbers dropped from ten in the first session to two in the following years and in

The tomb of Charles Piazzi Smyth and his wife
at Sharow churchyard, near Ripon in Yorkshire.

1855 he began limiting himself as he put it in the University Calendar 'to receiving any matriculated applicants for Practical Astronomy and advising or assisting such gentlemen afterwards in their studies, at various periods through the session'.

This was not what the University expected and in 1872 in the course of an enquiry into his Chair the University Court informed Piazzi Smyth that they considered his original and main salary of £300 p.a. to be attached to the Chair and they could not any longer acquiesce in his supervisory arrangements as being an adequate discharge of his duties as Professor. Piazzi Smyth's reaction was to put what he called 'these revolutionary and in spirit red-republican claims of the University Court' before his Board of Visitors with the proposal to separate his office as Astronomer Royal from that of a University Professor. He claimed that his main duty was to the observatory, and that, in any case, there was little point in continuing with a Chair of Astronomy since 'there are generally no students at all for so untoward, despised and poverty-stricken a subject in Scotland'.

On urgent representations from friends and colleagues Piazzi Smyth agreed to withdraw that unfortunate proposal, but when in May 1875 the University Court restated their view of the joint post and of the position of the Professorship within it Piazzi Smyth suggested to his Board of Visitors that the matter should be settled 'by the highest authority in the land' and that the whole subject of the Royal Observatory was now 'in a ripe state for thorough and searching legislative reform'.

Major changes were indeed to come, but not before Piazzi Smyth retired from his joint post in 1888 having completed in their entirety all the observational programmes of the Observatory. His resignation had been to a large extent forced on him. After 42 years in office it must have been an emotional moment for him when he signed his last entry in the Time Ball book 'C. Piazzi Smyth, Astronomer Royal for Scotland on this his last attendance at the Royal Observatory Edinburgh'.

He moved away from Edinburgh to a charming country house in Ripon in Yorkshire which he named 'Clova' in memory of the beautiful Glen of Clova in Aberdeenshire where his wife had been happy in her youth. He remained active to the end continuing his high-dispersion solar spectroscopy and recording many atmospheric phenomena. In 1896 he had the misfortune of losing his beloved wife who died after a long and painful illness. He himself died on 21st February 1900 at the age of eighty-one and was buried beside his wife in the churchyard of Sharow, a village some two miles away from Ripon. A pyramid marks their grave. They had no children, and the house 'Clova' went to his wife's brother, T. D. MacKenzie of Bombay. It was evacuated after Sir Duncan George MacKenzie's death in 1965 and is now a private nursing home.

The tablet which Piazzi Smyth had put on the pyramid in memory of his wife contains the following passage:

> In memory of Jessie Piazzi Smyth, the dear wife of Charles Piazzi Smyth, LL.D. Ed., late Astronomer Royal for Scotland, who was his faithful and sympathetic friend and companion through 40 years of varied scientific experiences by land and sea, abroad as well as at home, at 12000 feet up in the atmosphere of the wind-swept Peak of Teneriffe as well as underneath and upon the Great Pyramid of Egypt.

Following his own death Piazzi Smyth's family put a second tablet on the pyramid in which they described him 'as bold in enterprise as he was resolute in demanding a proper measure of public sympathy and support for astronomy in Scotland'.

With Piazzi Smyth died one of the most fascinating personalities in the scientific world of the Victorian era. Controversies particularly over his mystical interpretation of the Great Pyramid have done much damage to his reputation, but there can be no doubt that when he confined himself to science he was a man of the first order, a remarkable observer of the world around him and a great pioneer in many fields, particularly in spectroscopy. He possessed the gift of curiosity and inventiveness while being a meticulous and critical experimenter. He passionately believed that progress in science and particularly in his own field of astronomy could advance only in parallel with improvements in observational instruments, by working with the hands as well as with the head, with tools as well as on paper. In his ideas on setting up astronomical observatories on mountain sites and quite particularly in his objective and scientific manner of choosing the best possible location, he was far ahead of his time. It was a great pity for British astronomy that his recommendations were not heeded sooner, and that generations of British astronomers were hampered for want of worthy facilities in a good climate. Happily today Scotland's Royal Observatory maintains telescopes in a number of excellent sites abroad, and Piazzi Smyth's idea of the 'Travelling Astronomer' has become an everyday reality.

Part Two

The New Royal Observatory 1888–1957

6

RALPH COPELAND AND THE EARL OF CRAWFORD

IN 1888 when Charles Piazzi Smyth retired the fortunes of the Royal Observatory were at a very low ebb. The buildings were inadequate, the instruments poor and outmoded, the scientific publications stowed away for want of a library, the connection with the University strained. Who would have thought then that within a few years a splendid new edifice would be rising on the outskirts of the City, fully equipped, handsomely supported and enthusiastically welcomed by town, gown and Government? Yet this is what happened; and paradoxically it was the sadly neglected state of the institution, threatened with imminent closure, which inspired a patron to come forward with an extremely generous and immediately effective rescue plan.

The patron was Lord Lindsay, the 26th Earl of Crawford and Balcarres, Scotland's Senior Earl, himself an astronomer of distinction who possessed a private observatory at his home at Dunecht, 12 miles west of Aberdeen, which he had built with the assistance of his father in the 1870s and which he had equipped with first-class instruments of various kinds. Though occupied with his duties first as a member of Parliament and later, on succeeding his father to the Earldom, in the House of Lords, he had achieved a considerable reputation for himself as an astronomer, becoming a Fellow of the Royal Society of London in 1878 the same year in which he was elected President of the Royal Astronomical Society in succession to the illustrious Sir William Huggins. At the time of which we speak he was still only in his early forties, but by then the Dunecht Observatory had already acquired a world-wide reputation employing a staff of two professional astronomers and issuing its own publications.

Following a macabre episode in 1881 in which the body of his father was snatched from the Chapel of Dunecht House the young Earl had considered the possibility of moving his observatory from Dunecht to the family seat at Balcarres near Colinsburgh in Fife. He was at the same time very familiar with the problems of the Royal Observatory on Calton Hill. Having been the Chairman of two Government Commissions of Inquiry of 1876 and 1879 into the functions of the Royal Observatory he fully understood the difficulties facing its work. However, the Commissioners also saw that a very large outlay would be needed if the Observatory were to be brought to a proper level of efficiency and importance. They considered that this could not be done on the existing site and that nothing short of a complete removal of the observatory to a more favourable situation would solve the problem.

No action, however, was taken during Piazzi Smyth's tenure.

Meanwhile a Royal Commission on the Scottish Universities was putting recommendations before Parliament which included the question of the future of Astronomy in the University of Edinburgh. The proposal was that the Regius Professor of Astronomy should no longer be the Astronomer Royal for Scotland, that the institution of the Royal Observatory should be abolished and that the building on the Calton Hill should cease to be a national institution and should instead be handed over to the University.

The Earl of Crawford described afterwards his reaction on hearing this passage of the Bill when it was brought before the House of Lords. His first reaction was one of indignation, his second was one of sorrow and his third was that 'it shan't be!'. It was then in 1888 that he came forward with the remarkable proposal that he would present to the nation as an outright gift both the instruments of his observatory and his unique astronomical library of upwards of 15,000 printed books, pamphlets and manuscripts on the sole condition that the Government would build and maintain a new Royal Observatory to replace the old one on the Calton Hill.

This princely offer was accepted after some deliberation by the Government and a Committee with the Earl of Crawford in the Chair was appointed by the Secretary of State which was to select a suitable site for the new observatory and prepare plans for its building. Soon afterwards, on 28th January 1889, Ralph Copeland who had been Lord Crawford's chief astronomer at Dunecht since 1876 was appointed third Astronomer Royal for Scotland and fourth Regius Professor of Astronomy in the University of Edinburgh under exactly the same terms as those of his two predecessors.

Copeland who was aged 52 at the time of his appointment was a man of extensive intellectual attainments and considerable scientific experience. Having had few educational opportunities in his youth, his academic achievements were all the more remarkable, as they were entirely due to his own efforts and determination, coupled with a fine native intelligence. He was born on 3rd September 1837 on a farm at Woodplumpton, a village not far from Preston in Lancashire. His father, a farmer and part-time mill-owner, died when Ralph was only three years old. He had his first lessons from a local handloom weaver who ran a small school in his cottage, and later at the age of eight he was sent to the Grammar School at Kirkham. On leaving school he was apprenticed to his older brother who had a cotton mill in Blackburn. Life at the mill was not to the liking of Ralph who was of an adventurous nature, and at 16 years of age he set off for Australia in an emigrant ship to seek his fortune. It was at the height of the gold rush in Australia; gold had been discovered for the first time in the Victoria province in 1851 and people flocked there in their thousands from Europe as well as from Australia. For a while Copeland joined the search in the wild area of Omeo in the Australian Alps, but for most of his five years in Australia he worked as a shepherd for a Scottish sheep farmer in the same district.

It was the sight of the night sky in the Australian countryside which stimulated Copeland's interest in astronomy to the extent that he asked his mother to send him a small telescope, and with the help of this and books such as Sir John Herschel's *Outlines of Astronomy* which he had brought

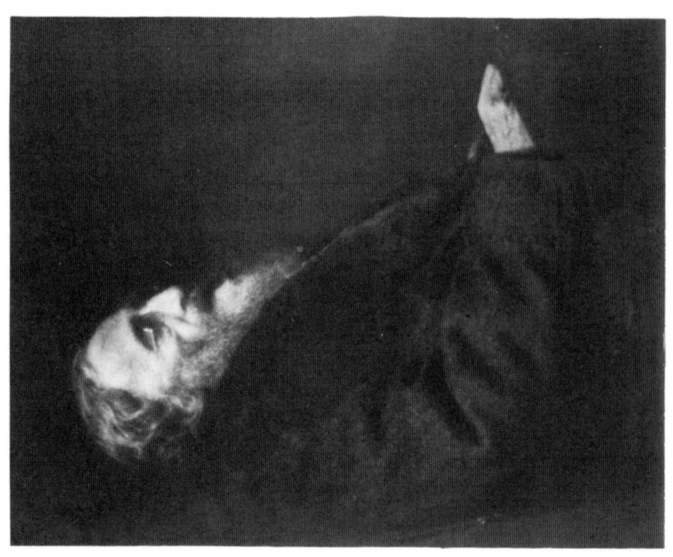

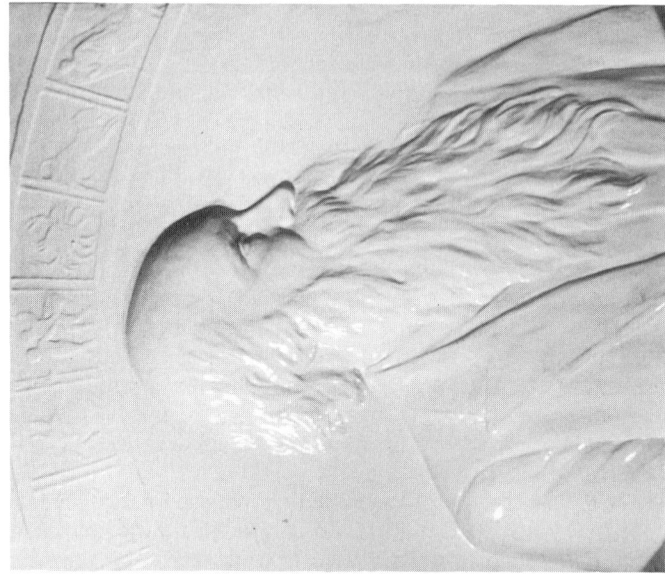

Left: James Ludovic Lindsay, 26th Earl of Crawford and Balcarres. *Right*: Ralph Copeland: a bas-relief in the entrance hall of the Observatory which was donated by the Copeland family after his death.

along – together with a Bible and a Shakespeare – he gained his first acquaintance with the heavens. He decided to return to England and, if possible, to enter Cambridge University to educate himself in the sciences. On his voyage home in the summer of 1858 he sailed around Cape Horn and during this voyage he was the first to observe the appearance and gradual development of Donati's famous comet, a sight which with a sea horizon must have been extremely impressive.

When Copeland on his return to England discovered that it was impossible for him, perhaps for financial reasons, to go to Cambridge, he joined the engineering works of Beyer, Peacock and Co. in Manchester as a volunteer apprentice while pursuing his mathematical studies privately. The engineering experience which he gained in this way was to stand him in good stead in later years, but his particular interest stayed with astronomy and he succeeded in setting up in conjunction with some fellow-apprentices a small observatory with a 5-inch Cooke refractor at a place near Manchester. Here in April 1863 he made his first published astronomical observations of lunar occultations.

In 1864 Copeland decided to give up his engineering work for a professional career in astronomy. To get the necessary academic training he chose the University of Göttingen in Germany where Professor Ernst Friedrich Wilhelm Klinkerfues, successor of the great mathematician and astronomer Karl Friedrich Gauss, had established a prominent school of astronomy and where physics was taught by Wilhelm Eduard Weber, famous for his work on electromagnetism and wave theory and the first to construct an electromagnetic telegraph which connected the Göttingen Physics Department with the Observatory. The German University system then prevailing of lecture courses and seminars which allowed students to present themselves for Doctorates without compulsory first degrees, was particularly suitable for able candidates like Copeland who was also of mature age. Nevertheless, the decision to throw up his job to become a student was a very courageous step for a man, by then 27, who had married in 1859 – to a first cousin, Susannah Milner – and who was the father of a little girl.

Copeland was at all times a good linguist, but determined not to be baulked in any way by language difficulties he spent eight months in a small village in Hesse to acquire a perfect command of German before he formally matriculated in the University of Göttingen in the spring of 1865. Amongst Copeland's fellow students at Göttingen was Carl Börgen, later the Director of the Observatory at Wilhelmshaven. Copeland and Börgen became close friends who between 1867 and 1869 working as assistants in the Göttingen Observatory carried out a major observational programme with the Reichenbach meridian circle which yielded the positions of more than 3000 stars near the celestial equator down to the ninth magnitude. Their observations formed the basis of the first *Göttingen Star Catalogue* which was published in 1869. The accomplishment of this work helped Copeland to recover from the personal loss which he suffered in the summer of 1866 when his wife died leaving him an infant son as well as a little daughter.

In 1869 both Copeland and Börgen took their degrees of Doctor of Philosophy, in Copeland's case with a thesis on the orbit of the binary Alpha

Centauri, the same star on which rests the fame of Thomas Henderson.

While completing their theses an exciting opportunity arose for the two friends which they grasped with enthusiasm. This was an invitation to join the German Arctic Expedition which was to be sent out in June of the same year to explore the east coast of Greenland as far to the north as possible. The task of the astronomers was to make geodetic observations for the measurement of an arc of meridian near the pole. The expedition, led by Captain Koldewey, consisted of two vessels of which the first, the *Germania*, with Copeland and Börgen aboard reached its destination and wintered in latitude 74°32′. The second ship, the *Hansa*, was less fortunate; it was crushed by ice and its crew drifted on an ice floe for seven months before they reached a Danish settlement on the opposite coast. The loss of the *Hansa* which was primarily a supply ship meant that the expedition could spend only one winter instead of the planned two on their voyage. During the winter Copeland and Börgen made regular geomagnetic, meteorological and auroral observations. Their geodetic work was delayed when Börgen was badly mauled by a bear, but they still managed to measure a base-line and travelling northwards by sledge to reach a latitude of 75°12′. Here they climbed the highest mountain on Greenland, Petermanns Bjerg, at 9700 feet. The scientific results of this Arctic Expedition including Copeland and Börgen's geodetic work were published in 1874 in the second volume of the official account *Die Zweite Deutsche Nordpolarfahrt*. Names on the map of Greenland like 'Germania Land' or 'Koldewey Store' are reminders of the expedition which having lasted more than a year returned to Germany amid much acclaim, Copeland being one of those decorated by the Emperor William I. For Copeland it was the first of many scientific expeditions in the course of the succeeding thirty years.

Copeland's years in Germany were well worth the effort, and when he returned home at the age of 33 he was amply equipped for the astronomical life he had set his heart on as a boy. At Göttingen he had assimilated the latest contemporary ideas in physics through Weber and others and had acquired also a high level of mathematical training. From Klinkerfues, a recognised authority, he learned the traditional skills of positional astronomy and inherited a special interest in comets which he was to show in later researches. The Arctic Expedition introduced him to geophysics and terrestrial magnetism, adding to the broad outlook on science for which he became known. He also came away with a great personal attachment to his German colleagues who were to remain close friends through all his life. A further link with Germany was his marriage, a year later, to his second wife, Theodora Benfey, the daughter of Professor Theodor Benfey, a distinguished oriental scholar and the founder of the famous school of linguistics of Göttingen University. With her, he was to have another son and two daughters.

In 1871 Copeland went to Ireland to take up an appointment as assistant to the Earl of Rosse, who had his own private observatory at his home at Birr Castle. His father, the third Earl, had succeeded in 1845 in constructing a reflecting telescope which for many years was the largest in the world having a speculum mirror of 72-inches diameter. Copeland used this telescope for the observation of galaxies or what were then called nebulae. Most of his time was spent, however, with a smaller 36-inch telescope with which he measured in conjunction with Lord Rosse the thermal radiation of the Moon. These observations were a follow-up of Piazzi Smyth's work in Tenerife and were aimed at deciding whether the Moon's heat came entirely from the Sun's radiation or whether some of it was attributable to its own interior heat. The method was to measure the variation in lunar thermal radiation with lunar phase, and the result, published by Lord Rosse, was that the radiation is effectively solar in origin.

While at Birr Copeland whose fame as an arctic explorer had spread, was offered joint command of an Austro-Hungarian polar expedition in 1872. Though this must have been a tempting offer Copeland, fortunately for astronomy, declined. The actual expedition leader was Payer who had been an officer on the *Germania* during the earlier polar voyage. This long gruelling expedition culminated in the discovery of new territory at latitude 80° which was named Franz Joseph Land, and established Payer's position in the history of Arctic exploration. Copeland's opportunity for an exciting and purely astronomical expedition came before long. Shortly after moving from Birr to the Dunsink Observatory of Trinity College Dublin in 1874 Copeland was invited by Lord Lindsay to accompany him to Mauritius for the observation of the 'Transit of Venus' across the Sun which took place in December of that year.

Observations of transits of one of the inner planets, Venus or Mercury, across the disc of the Sun were of the highest importance at that time as means of determining a fundamental astronomical quantity, the Astronomical Unit, or the distance between the Sun and the Earth. As the planet Venus moves between the Sun and the Earth its path seen projected on the Sun's disk depends on the observer's location on the Earth, and observations of the phenomenon from two sites, widely separated in latitude, are capable of providing the absolute distances between the Earth, the Sun and Venus in terms of the length of a terrestrial base line. Hence the need for expeditions to the southern hemisphere. Transits are not common; Venus crosses the Sun only at intervals of about a century when two transits occur separated by only eight years. A pair of such transits occurred in 1874 and 1882; the next two will be seen in 2004 and 2012.

Both the 1874 and 1882 transits were occasions for expeditions in which Copeland took part. In 1874 he joined Lord Lindsay on his yacht *Venus* which on its outward voyage to the Indian Ocean called at the uninhabited Brazilian island of Trindade in the South Atlantic where Copeland made a botanical discovery finding a great tree-fern which is called after him – *Cyathea Copelandi*.

The transit observations at Mauritius which had been prepared with extreme care were unfortunately spoiled by cloud, but Copeland's association with Lord Lindsay was to have a decisive influence on the rest of his life. After a short stay at the Dunsink Observatory on his return from Mauritius Copeland was invited by Lord Lindsay in 1876 to join him at his Dunecht Observatory in succession to David (later Sir David) Gill who three years later was to become HM Astronomer at the Cape. All accounts agree that the next twelve years were a particularly happy time for Copeland. The remarkable equipment at Dunecht included a 15-inch refractor

The Observatory House at Dunecht where Copeland lived and worked from 1876 to 1889. The photograph was taken in 1882.

which was Copeland's principal telescope. There was also a 24-inch reflector and a great variety of spectroscopes and other auxiliary apparatus.

Thanks to the fact that Lord Lindsay was an enthusiastic bibliophile, Dunecht Observatory could also boast a large library, which included numerous rare volumes of old astronomical and mathematical books. Copeland took a great personal interest in this library which made him an ideal assistant to Lord Lindsay in his efforts to build up a truly unique collection.

Much of Copeland's time in his first few years at Dunecht was spent on the reduction of geodetic observations made in the course of the expedition to Mauritius. The results were published in the third substantial volume of the *Dunecht Observatory Publications*. His observational work at Dunecht was carried out chiefly with the 15-inch refractor which he used to determine the positions, motions and particularly the spectra of a number of bright comets. Observing with his assistant J. G. Lohse – who was later to join the staff of the new Astrophysical Observatory in Potsdam – he noticed the sudden appearance of the sodium D-lines in the spectrum of the Wells comet of 1882 two weeks before its perihelion passage. In observations of the spectrum of the great comet of the same year, the brightest comet of the 19th century, Copeland and Lohse discovered on 19th September in full daylight six bright lines in the green and yellow regions of the spectrum which were later identified with iron lines. At the time this was a significant discovery in the field of astronomical spectroscopy. These and other cometary observations were published by Copeland in the *Dunecht Circulars* and later in the international astronomical journal *Copernicus* which he edited jointly until its discontinuance in 1884 with his friend Dr J. L. E. Dreyer, the Director of Armagh Observatory and the compiler of the famous *New General Catalogue* (NGC) of star clusters and nebulae. *Copernicus* was a beautifully produced journal which attracted contributors from leading astronomers of the day and particularly from the continent of Europe. It was printed in Dublin with the financial backing of Lord Lindsay.

Apart from those of comets Copeland observed the spectra of variable stars and novae such as those which became visible in Cygnus in 1876 and in Andromeda in 1885. Undoubtedly his greatest spectroscopic discovery was made in 1886 when, using a new powerful Cooke spectroscope he noticed the D_3-line of helium near the sodium D-lines in the spectrum of the Orion nebula. At that time helium, the 'Sun element' had not yet been discovered on earth; it had been seen only in the spectra of solar prominences and the solar chromosphere. Copeland's observation proved that helium was not exclusively of solar origin, a demonstration that caused widespread interest and discussion.

Copeland's spectroscopic work suffered an interruption when in 1882 he was asked by Lord Lindsay to take part in the observations of the second of the 19th century pair of transits of Venus for a new determination of the solar distance. This time he went to Jamaica where, the weather being favourable, he was successful with his actual observations. His was one of a number of British expeditions which were scattered over the globe from Queensland to Bermuda and which had for the most part been fortunate with the weather. However, when it came to the discussion of the results of these expeditions and of those from other countries, some fifty in all, it became clear that it was too difficult to determine the critical moments of contact between the limbs of the Sun and Venus to allow the solar distance to be measured with the expected high precision. The widely differing results of all these efforts were published in 1887 in the *Monthly Notices of the Royal Astronomical Society*. It was to be more than fifty years before a substantial improvement in the value of the solar distance was to be achieved; this was based on observations of the close approach of the minor planet Eros in 1930–31.

Being in Jamaica and relatively not too far from South America Copeland decided to repeat Piazzi Smyth's experiments in 'Mountain Astronomy' by making certain astronomical tests from the slopes of the high Andes. On his journey which was eventful and adventurous he crossed the isthmus of Panama and sailed down the west coast of South America. He first tried to get to Quito in Ecuador, but was prevented by revolution from travelling any further than to its port Guayaquil. He then decided to travel south to Arequipa and to La Paz in Bolivia, establishing his headquarters at Puno in Peru on Lake Titicaca at an altitude of 12,500 feet. Here he made over a period of six weeks a number of important spectroscopic observations using a 6-inch refractor and a new three-prism Vogel spectroscope, called after the designer and at the time a very advanced astronomical tool. By scanning the sky he readily distinguished objects with emission spectra such as planetary nebulae and Wolf-Rayet stars. At Arequipa, La Paz and Vincocaya (now Sumbay), the highest point on his itinerary (14,360 feet) he made observations of the visibility of stars and planets in daylight to test the transparency of the sky and the resolvability of double stars to test the image definition or 'seeing'. He also took meteorological readings which he compared with the earlier observations of the great British mountaineer Edward Whymper, best known as the first conqueror of the Matterhorn. His conclusion was that the regions of Arequipa and La Paz were eminently suitable for 'mountain astronomy' and in fact, some years later the Harvard Observatory established a station in Arequipa while in 1926 the Potsdam Astrophysical Observatory did likewise on a site near La Paz. The third volume of the journal *Copernicus* reports on all these observations and also on a journey to the major North American observatories which he made on his way home in the summer of 1883. This marathon tour had occupied almost a whole year.

Following this expedition Copeland gave some thought to the possibility of moving the Dunecht Observatory, or at least some of its instruments, to a site in Jamaica, but events unforeseen by him at the time were to keep him and the observatory in Scotland.

Another expedition was in store for Copeland a few years later when he was asked by Lord Crawford to go to Central Russia to observe the total eclipse of the Sun in August 1887. The expedition was unsuccessful on account of bad weather, a disappointment which eclipse observers learn to endure. In this particular instance it was a pity that the expedition had taken Copeland away from what was to be his last spectroscopic work at Dunecht. This was started with the setting up in 1887 of an observing station on the top of the nearby Barmekin Hill where a large spectroscope with a Rowland grating was mounted complete with recording facilities for an investigation of the solar spectrum at medium and low altitudes. During Copeland's absence in Russia this important solar spectroscopic work was completed by Dr L. Becker who had succeeded Dr Lohse as Copeland's assistant and who later became Professor of Astronomy in Glasgow. It was published in Volume 36 of the *Transactions of the Royal Society of Edinburgh* and gained the Society's Makdougall-Brisbane Prize for Dr Becker, on Copeland's recommendation.

One of Copeland's very last tasks at Dunecht was the completion and passing through the press of the great *Catalogue of the Crawford Library* which when it was published in 1890 contained the titles of some eleven thousand books and of many pamphlets and manuscripts. The story of this Library will be described later in a special chapter.

It was most appropriate that Copeland should be chosen to look after the realisation of the proposed new observatory; indeed, though there was at least one other candidate, Dreyer, Copeland's appointment in 1889 seems to have been a foregone conclusion. It ensured continuity of the Dunecht work, and in this connection it gave Copeland particular pleasure to receive a letter from Piazzi Smyth written 'in the most friendly spirit' hoping that he would not be separated from the Dunecht instruments. Two of the Dunecht staff, Dr Halm and Mr J. McPherson, the engineer, joined Copeland in Edinburgh, while the other astronomer, Dr Becker, remained at Dunecht until the removal of the instruments.

Thus what was effectively the amalagamation of the Dunecht and Calton Hill Observatories marked the end of Dunecht which in the short time of its existence, had been a remarkably happy and successful institution. Sir David Gill, Copeland's first and only predecessor at Dunecht, who was at this time HM Astronomer at the Cape, wrote a charming letter to Copeland congratulating him on his appointment while at the same time looking back nostalgically on his own Dunecht days:

You will have a most desirable position in every way – except perhaps in clearness of sky – the most delightful society – and an equipment second to none in Great Britain. I can hardly think without a sigh of those foundations at Dun Echt which I laboured to make so satisfactory and sound – all swept away – and yet I am sure it is for the best interests of science that it should be so. You have a grand chance, with all your experience, to plan a splendid observatory – and I am sure you will.

7

THE MOVE TO BLACKFORD HILL

IT NOW became Copeland's first duty to select a suitable site for the proposed new observatory and to consider plans for its construction. He also had to give thought to the preparation of a course of lectures to students of astronomy at the University which had lapsed many years before. In the meantime astronomical work was to be continued at Calton Hill and at Dunecht until such time as the new observatory was ready to receive the instruments from both places.

As regards the choice of site, Copeland was assisted by Lord Crawford's Committee which consisted of Lord McLaren of the Court of Session, himself a well-known amateur astronomer, Professor Peter Guthrie Tait of the Chair of Natural Philosophy in the University of Edinburgh as well as Copeland himself. The Committee considered various locations at moderate distances from Edinburgh including Dunsapie Hill, Craigmillar Castle, the Forth Bridge area, the Haddington and Pentland Hills, the Braids and Blackford Hill. The Haddington and Pentland areas though meteorologically suitable were ruled out as too distant. The south side of the City was favoured because of the prevailing southerly winds which would discourage pollution by the City's smoke. Following extended tests of the stability of the rock foundations and of meteorological factors carried out in the autumn of 1889 by Dr L. Becker, Copeland's assistant from Dunecht, Blackford Hill was chosen unanimously as being most suitable on all counts with the added advantage of being a public park 'not liable to be seriously encroached on by future buildings', and yet being not too far away from the University. The ground of the Blackford Hill belonged to the City of Edinburgh, but on due representation to the Town Council the desired area was liberally ceded to the Government on the sole condition that it should be used only for the purposes of a national observatory. At the same time discussions were held on the transfer of the buildings and some of the instruments of the Calton Hill observatory to the City of Edinburgh on the completion of the observatory buildings on the Blackford Hill.

Following enquiries by Copeland about the cost of some recently erected observatories abroad such as those at Vienna and Strasbourg plans for the new buildings on Blackford Hill were prepared by Mr W. W. Robertson,

The north face of the Royal Observatory on Blackford Hill, photographed when building operations were completed.

Chief Architect to HM Office of Works in Scotland, in close consultation with Lord Crawford's Committee. These plans were approved by HM Treasury in 1891 at a cost in excess of £30,000, considerably higher than early tentative estimates. Though on one occasion the Treasury had suggested that certain economies might be made, in fact the entire project went through as planned and there was never any real difficulty regarding financial support. As Copeland had pointed out before he took office, Lord Crawford's gift was worth five times that of any building which would house it.

The actual construction of the observatory and staff dwelling houses was undertaken by Messrs W. and J. Kirkwood of Edinburgh who in 1892 started their work by building an inclined railway from the eastern end of the observatory grounds to a temporary siding at Blackford railway station below. Amongst the material taken up the hill by this railway was the reddish sandstone from Doddington Hill in Northumberland which the architect decided to use for the facing masonry. Building progress was fast, and by the summer of 1893 a contract could be signed with Sir Howard Grubb of Dublin for the construction of the two revolving domes of 38 and 22 feet in diameter at the east and west ends of the building which were to cover the two principal telescopes of the observatory, the 15-inch refractor and a 24-inch reflector. Copeland chose the cylindrical drum form of covering in preference to the more usual hemispherical shape of dome because he considered it as better suited to the Scottish climate. The year 1893 also saw the preparations for the transfer from Dunecht of both the 15-inch refractor and the 8.6-inch transit circle.

In 1895 the general move was largely complete and Copeland was able to leave the observatory's house at 15 Royal Terrace and move into his spacious new residence on Blackford Hill. The two assistants' houses were soon afterwards occupied by Mr Heath from Calton Hill and Dr Halm from Dunecht. The very first to reside on Blackford Hill was in fact the engineer J. McPherson who coming from Dunecht moved with his family into the lodge.

By the beginning of 1896 all telescopes were in position and the new Royal Observatory was ready to be formally opened on 7th April by the Secretary of State for Scotland, Lord Balfour of Burleigh, in the presence of Lord Crawford and a large congregation of distinguished guests. The company included the Principal of the University, Sir William Muir, many Professors, the Lord Provost and other prominent citizens of Edinburgh. In his address on that occasion Copeland paid a particularly warm tribute to the work of his predecessor Piazzi Smyth who had sent his best wishes being himself by that time not well enough to attend. The new observatory attracted naturally great public interest as is evident from the records of the numerous scientific and other parties who came to visit it in the following years.

Copeland had every reason to be proud of the new observatory. Lord Crawford's Committee had desired that it should be a stately building which would do justice to its commanding position overlooking the City of Edinburgh, and indeed it was one of the finest Victorian buildings in the city and a much admired feature of the Edinburgh skyline. A contemporary account

Top: An early photograph of the Royal Observatory on Blackford Hill showing the library wing and entrance porch. *Bottom:* The Meridian House on the west end of the Royal Observatory, Blackford Hill, now replaced by modern buildings.

describes its architecture as 'Italian in style with a free admixture of Greek feeling' and Brian Crosland in his book on Victorian Edinburgh judges it an outstanding example of the 19th century Italianate Revival.

The main building, 180 feet long, is oriented east to west with octagonal towers at each end surmounted by the cylindrical telescope housings. These are covered with sheets of copper which with their green weathered colour blend beautifully with the red sandstone masonry. The exterior decoration of the observatory is replete with astronomical references. The long north wall of the main building overlooking the city has a row of eleven high windows, alternating with twelve panels carved with the signs of the zodiac. On the pediment of the north facing window of the smaller tower is a beautiful carved representation of Phoebus, the Sun god, driving his chariot, while on the south wall of the same tower is a sundial flanked by Dawn and Dusk and carrying the legend *Docet Umbra*. The Library wing extending southwards from the main building has a pillared entrance with the Scottish coat-of-arms and, carved above the door, the Psalmist's words *Coeli enarrant gloriam Dei*. According to the fashion of the day cornices, walls and copper drums are lavishly embellished with flowers, gargoyles and other emblems. There are numerous reminders of the Royal connection in the form of Queen Victoria's monogram and the intertwined R O initials.

The south facade of the Library carries a medallion of Lord Crawford and the date 1894, when the actual building work was completed. The same wing has inscribed along the cornice the names of great astronomers of the past – Copernicus, Galileo, Kepler, Newton, Herschel, Bessel and Scotland's first Astronomer Royal Henderson. This gallery of fame brings to mind the inscriptions on the walls of the legendary observatory of Uraniborg on the Baltic island of Hveen, built for the great 16th-century Danish astronomer Tycho Brahe. Copeland was an ardent admirer of Tycho Brahe's work and had inscribed in the form of a scroll over the entrance to his new residence Tycho's motto *Nec fasces nec opes sed sola artis sceptra perennant* (it is neither honours nor riches, but only the achievements of the arts which will endure). It was typical of him that before deciding on the lay-out of the Latin words on the scroll he consulted his colleague Professor H. C. Goodhart of the Chair of Humanity. At the opening ceremony on Blackford Hill the company could also see hanging on the wall of Copeland's study a splendid portrait of Tycho Brahe which had been acquired for the Crawford collection together with one of James Short, the 18th-century Edinburgh telescope maker.

There was provision on the first floor of the main building for a solar laboratory and the flat roof joining the two main turrets was intended as an observing platform for smaller instruments and for practical instruction of students. The Dunecht transit circle was housed in a separate meridian house to the west of the main building. Magnificent as the exterior of the observatory was, its interior was also most carefully planned by Copeland to contain every possible accessory for what he called 'physical astronomy' which meant largely optical and spectroscopic work. The laboratories included even a magnet room where a powerful magnet could be used for impressive demonstrations. All in all, the new Royal Observatory standing in beautiful landscaped grounds was a splendid and worthy repository for the Crawford gift and could hold its own among the national observatories of the world.

By the time that Copeland was able to move up to Blackford Hill he had seven very strenuous years behind him. While he was kept busy over the construction of the new observatory he travelled regularly to Dunecht, some 120 miles from Edinburgh, to continue his astronomical work while

Top: A representation of Phoebus the Sungod and his chariot carved on the north pediment of the West Tower of the Observatory. *Bottom:* The sundial with the figures of Dawn and Dusk on the south face of the Observatory.

the instruments remained there, usually calling on the way at Balcarres in Fife to report progress to the Earl of Crawford. There were many journeys to London, York and Dublin to discuss equipment for the new observatory. At Dunecht Copeland and Becker continued their spectroscopic observations of planets, comets and nebulae. They observed the spectrum of Nova Aurigae when it was discovered in 1892 by an Edinburgh amateur astronomer, the Rev. Dr T. D. Anderson who announced his discovery on an unsigned postcard to Copeland five days after he had first seen the Nova.

Dr Becker who remained at Dunecht until his appointment in January 1893 to the Glasgow Chair of Astronomy was able to complete at Dunecht a series of positional observations with the transit circle of some two hundred nebulae. These observations were published in 1902 as the first of four papers which together made up the first volume of the new *Annals of the Royal Observatory Edinburgh*.

Copeland also tried to make use of the transition period before the completion of the new observatory by embarking on an independent reduction of the old positional observations which had been made on the Calton Hill by Henderson and Wallace. In connection with this programme in which he wanted to use the most recent improved values for the fundamental astronomical constants he paid a visit in 1893 to Professor A. Auwers in the Prussian Academy of Science in Berlin who was the undisputed Dean in the field of positional astronomy at the time. Auwers had just had an honorary degree of Doctor of Laws conferred on him by the University of Edinburgh, but had been prevented by ill health from attending the ceremony.

While in Germany, Copeland took the opportunity of spending some time with his old friend and fellow student and explorer, Carl Börgen, at Wilhelmshaven, and of paying a visit to Captain Koldewey, master of the Greenland expedition of 24 years earlier, who was now in charge of an oceanographic laboratory in Hamburg and engaged in developing magnetic equipment for the Antarctic. Copeland also visited a number of observatories and the works of the famous instrument makers J. G. Repsold in Hamburg.

Copeland's ambitions for the re-reduction of the Henderson-Wallace catalogue turned out to involve considerably more labour than originally anticipated and its result 'A New Reduction of Henderson's catalogue for the Epoch 1840.0' was ready only shortly before Copeland's death. In fact, at the time of his death Copeland was still working on the introduction to this volume which was then completed by Dr Halm and printed in 1906 as Volume II of the *Annals of the Royal Observatory*.

Throughout all these years Copeland never forgot the Professorial duties of his post and by contrast with his predecessor took a very active interest in the affairs of the University. The routine work of the observatory also continued without interruption during the transition phase. The duties consisted first and foremost in the provision of a regular time service. When this activity had to be transferred from the Calton to the Blackford Hill the difference of longitude between the two observatories had to be determined; the new observatory was found to be 1.17 seconds west of the old one. As on Calton Hill the new observatory provided time by telegraph wire for the firing of a gun at Edinburgh Castle (for which service a sum of £50 for gunpowder was included in the observatory's budget!), the dropping of a time-ball on the Nelson Monument on Calton Hill, for the firing of a time gun in Dundee and for the correction of a clock in the naval Dockyard in Rosyth. The observatory also controlled clocks in the University and at the GPO. The Dundee time service was discontinued at the time of the First World War and was never resumed. The Edinburgh one o'clock gun was only once suspended for a short period at the end of the war in 1918 because of its distressing effect on shell-shocked soldiers.

At the old Royal Observatory meteorological duties had included the coordination of observations from 55 stations throughout Scotland for which the observatory had received a subvention of £100. Piazzi Smyth had always complained bitterly about this chore, and it was only when Copeland took over that a sensible arrangement was reached with the meteorological Society (now the Meteorological Office) whereby the Society took on the responsibility of collecting observations and the Royal Observatory became – and still is – one of the local meteorological stations making daily readings of temperature, pressure, wind velocity, rainfall and hours of sunshine.

Seismological records were also kept up regularly, and in December 1900 Copeland replaced the original bifilar pendulum seismometer with a Milne-Shaw seismograph which produced a valuable series of seismograms until it was replaced by more modern instrumentation in the early 1960s.

The building of the new Royal Observatory had been a remarkable achievement, but Copeland's love of travel and adventure did not diminish with the completion of that work. No sooner were the celebrations surrounding the opening of the observatory behind him than Copeland plunged into the preparations for the first of three solar eclipse expeditions which he was to lead from Blackford Hill. The location of the eclipse due to take place in August 1896 was Vadsö, a remote spot on the east coast of Finmark in northern Norway, close to the Finnish-Russian border. For two months Copeland and two of his staff were occupied preparing and testing the instruments for the eclipse, including a large horizontal telescope and two equatorials with objective prisms with which the spectrum of the solar corona was to be photographed. Forty-five hundredweight of equipment was shipped out from Leith a month in advance and a few days later the expedition party of four including Copeland's younger son Theodore who took part as a volunteer sailed from Newcastle to Stavanger. Unfortunately, all preparations came to nothing when the sky was completely cloudy during the total phase of the eclipse.

However, the disappointment was to some extent softened when on the very day of Copeland's departure from Vadsö the explorer Fridtjof Nansen returned in the *Windward* to the neighbouring port of Vardö from his epic three-year Arctic Expedition which had taken him to within 4 degrees of the North Pole. Within hours of his arrival he and Copeland met and their acquaintance was renewed in the following year when Nansen and his wife were Copeland's guests at the Royal Observatory. Polar exploration was always one of Copeland's special interests and he was one of those who supported from the beginning the idea of a Scottish Antarctic Expedition.

In 1897 Copeland was asked by the Joint Permanent Eclipse Committee

Top: Copeland with his 40-foot horizontal telescope and coelostat at the total solar eclipse of January 1898 at Ghoglee near Nagpur, India. *Bottom:* The solar corona observed by Copeland in India in 1898 (Royal Observatory Edinburgh).

of the Royal Society and the Royal Astronomical Society which had been set up five years earlier to coordinate eclipse expeditions, to take part in the observations of a solar eclipse occurring in India in January 1898. Copeland could not resist the call and this time he was entirely successful. Observing from a station near Nagpur in the Central Provinces he used the same instruments which he had employed in Norway, a horizontal telescope nearly 40 feet in focal length, a second shorter photographic telescope and a prismatic camera with quartz optics. From this expedition Copeland arrived home from India a month later – and resumed his lecturing at the University on the following day. A preliminary account of his results was later printed in the *Proceedings of the Royal Society* and also as an Appendix to Volume 58 of the *Monthly Notices of the Royal Astronomical Society*.

Two years later Copeland was again asked by the Joint Permanent Eclipse Committee to organise an eclipse expedition, this time to Spain for the observation of the eclipse of May 1900. Once again he agreed and once again he was successful observing from Santa Pola near Alicante with the same equipment he had used before. He was accompanied by Mr Heath, his first assistant, and also by Franklin-Adams, an amateur astronomer, now well known for the photographic charts of the whole sky which he produced working first in England and later in South Africa. The expedition led to photographs of solar prominences and of the corona and of spectrograms of the solar chromosphere which were published in an Appendix to Volume 60 of the *Monthly Notices of the Royal Astronomical Society* and in the *Proceedings* of both the Royal Societies of London and Edinburgh. Copeland was particularly pleased with the photographs he had obtained of the Indian eclipse and he asked Mr W. H. Wesley, Assistant Secretary of the Royal Astronomical Society and a well-known expert on the interpretation of celestial photographs, to produce a drawing of the corona from the best of the photographs. The delivery of this drawing late in 1904 when Copeland was already in very poor health, brought him the greatest pleasure.

In 1898 a major observational programme of stellar positions was started with the Transit Circle on Blackford Hill. Apart from its use for the time service this was the last programme for which the instrument was employed. The aim was the determination of the positions of some 2700 stars near the ecliptic which had been selected by Sir David Gill of the Cape Observatory as suitable reference stars against which the movements of the major planets could be measured with high accuracy leading to an improvement in the determination of their orbits. The results of this valuable work were published only in 1910 after Copeland's death by his successor as Volume III of the *Annals of the Royal Observatory*.

The 15-inch refractor was used as at Dunecht for the observation of minor planets and comets and for spectroscopic studies of the bright planets and nebulae. In December 1900 a violent gale caused considerable damage to the eastern dome of the observatory in which this telescope was housed, the only serious damage ever suffered by these domes. The revolving drum was dislodged from its wheels and it was not until the last day of the year that it was lifted back into its original position. Copeland reports that by the 20th February the observations with the 15-inch refractor could be resumed with safety – just in time for the observation of the first major astronomical phenomenon of the new century, the appearance of Nova Persei which was discovered on 22nd February by the same Dr T. D. Anderson who in 1892 had discovered Nova Aurigae. Using the 15-inch refractor with a Cooke spectroscope attached to it Copeland and Halm were able to make a series of observations of the spectrum of the Nova and its changes in the early stages

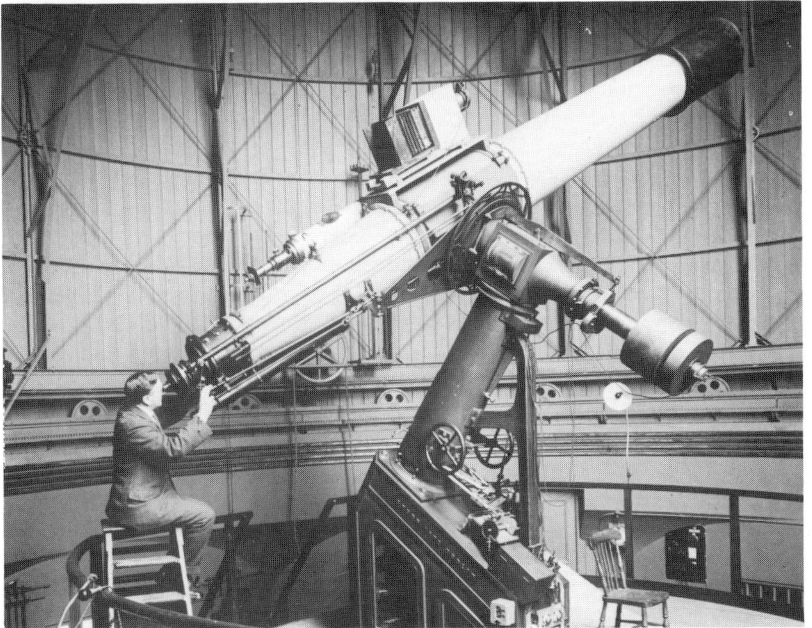

Top: Copeland at work with his favourite instrument, the spectroscope attached to the 15-inch telescope. *Bottom:* The 15-inch telescope in the East Tower of the Royal Observatory.

of its development.

The appearance of the Nova had been announced by Copeland in No. 54 of the *Edinburgh Circulars*, the last Circular he was able to issue. Shortly afterwards he fell seriously ill, first in the summer of 1901 with influenza and in the following years with a series of attacks of angina pectoris. Though he never regained his former health he managed to carry on with much of his scientific work. He handed over the non-scientific administration of the Observatory to his first assistant, Mr Heath, at the end of 1903, but he continued his University activities relinquishing only the advanced course to Dr Halm. A year later, however, he was obliged to leave all the teaching to Dr Halm who was then appointed Pro-Professor by the University. All the while Copeland insisted on continuing to issue his Annual Reports the last of which appeared in April 1905, six months before his death.

Copeland's Edinburgh years had been dauntingly active – observing, teaching, travelling and advising and receiving visitors all of which he did with tremendous enthusiasm and generosity. At home he became a much admired figure. When in 1903 the appearance of a new comet (comet 1903c) was announced, which was in fact unfavourably placed for observation from Scotland, Copeland had a request for information from a member of the Royal Household at Balmoral. His reply contained a lengthy explanation of cometary orbits and the practical problems of observing the comet in question: 'Needless to say, I shall be most ready to forward to Balmoral a suitable telescope for the use of Her Majesty (i.e. Queen Alexandra), but I regret to repeat that even with the aid of a practical observer it will now be almost impossible to obtain a view of the comet'.

Many of the leading British scientists of the day, astronomers and others, came to Edinburgh to visit Copeland and inspect the new Observatory. Even while the Observatory was still under construction he enjoyed taking his visitors up to Blackford Hill to see the building operations. When Mr E. W. Maunder, the Greenwich solar observer – of 'Butterfly-diagram' fame – came to look at these, he expressed the hope that the Royal Observatory at Greenwich would soon follow Edinburgh's example and move out into the country. Copeland endorsed this, adding that a new site could well be on the same zero meridian. It is interesting that Copeland thought it worth recording his conversation with Maunder which took place more than half a century before the Royal Observatory actually moved from Greenwich to Herstmonceux in Sussex, though not to the same meridian.

A large number of visitors came from the Continent of Europe and from the United States where Copeland had made many contacts on his tour in 1883. Amongst the latter were E. E. Barnard and W. W. Campbell who came from the Lick Observatory in California and E. C. Pickering from Harvard. Copeland's common interest with Campbell and Pickering was stellar spectroscopy and the possibility of entering the field of radial velocity measurement.

Copeland was deeply impressed by Barnard's photographic work and in particular by his beautiful photographs of the Milky Way. The potential of the photographic plate struck him particularly when it was demonstrated to him, a regular observer of comets, that he could follow these photographically to much greater distances than he was able to do visually. While

Barnard was in Edinburgh Copeland was asked to present to him formally the French Lalande Gold Medal for 1893.

Copeland's correspondence reveals how widely known and greatly respected he was both nationally and internationally. The Arctic experience of his early years was often called on in connection with scientific programmes for polar expeditions and so were his experiences of mountain astronomy in the Andes which led to a correspondence with a number of astronomers including Dr George Ellery Hale, then at the Yerkes Observatory at Williams Bay, Wisconsin, and later the famous Director of the Mount Wilson Observatory in California. The correspondence centred on the question of setting up high altitude observatories where Copeland took the view that for any permanent establishment the human factor ought not to be neglected and extreme conditions at excessive altitudes should be avoided.

In his world-wide contacts, Copeland was greatly helped by his mastery of languages; he could speak and write French and German with the same facility as English. In his last months, when prevented by ill health from active work, he occupied himself by studying Persian and, according to his close friend Dr Dreyer, he was at the time of his death fully able to read with pleasure the writings of Omar Khayaam in the original.

Copeland, in short, was in his life-time well known far and wide as a personality and as an astronomer. For his astronomical successors, however, his most important contribution was that he made the new Royal Observatory Edinburgh internationally known, an institution which could hold its own amongst the world's great observatories, as the science of Astronomy entered the 20th century.

In his declining years, Copeland contemplated retiring, but found that he could not afford it since his relatively short service entitled him to very little Government pension. After Copeland's death Lord Crawford in a letter to *The Times* protested strongly at what he considered a serious lack of provision in such cases.

Copeland passed away on 27th October 1905 in his residence on Blackford Hill surrounded by his family. One of his daughters, Fanny (Mrs Copeland Barkworth), who had been trained as a musician at the Berlin Academy of Music, always retained a close interest in the affairs of the Royal Observatory. Like her father for whom she had the very greatest admiration, Fanny was herself to lead a life full of adventure which took her as Reader in English to Llubjana University in Jugoslavia where she died having become a national figure there in her 98th year in 1968.

8

FRANK DYSON AND GREENWICH

FRANK WATSON DYSON, who at the early age of 37 succeeded Ralph Copeland, had the distinction – so far unique – of being successively Astronomer Royal for Scotland and Astronomer Royal at Greenwich, the fourth Astronomer Royal for Scotland and ninth Astronomer Royal. His appointment followed very much the tradition south of the border of recruiting to the highest astronomical posts scholars trained in the University of Cambridge. It was also the first ever interchange between the two Royal Observatories, Dyson being at the time of his appointment to Edinburgh Chief Assistant at Greenwich. The short spell, less than two months, which elapsed between Copeland's death and the choice of his successor, indicates that there can have been no hesitation on the part of the Secretary of State for Scotland or the Court of the University of Edinburgh in recognising Dyson's outstanding suitability for the post.

Dyson, the son of a Baptist minister, the Rev. Watson Dyson, was born at the Manse in Measham in Leicestershire on 8th January 1868, the eldest son of a family of seven. Much of his early life was spent in Halifax and he regarded himself all his life as a Yorkshire man. His mathematical abilities became clear when at the age of thirteen he won a junior certificate in the Cambridge local examination in which he took first place in all England. As a result Frank was offered a Governors' scholarship at Bradford Grammar School which he entered in 1882 under its remarkable headmaster, the Rev. William Keeling. Four years later at the age of eighteen he won an open scholarship to Trinity College, Cambridge, where he excelled in mathematical studies. After his first two years he was awarded the Sheepshanks Exhibition in Astronomy and he became second wrangler in the Mathematical Tripos of 1889. As a postgraduate student Dyson concerned himself with intricate problems of theoretical mechanics, becoming Smith's Prizeman in 1891 and Fellow of his College in the following year. He also gained the Isaac Newton Studentship, a well-known door to an astronomical career.

Dyson did not remain long in Cambridge. In 1894 he was offered the post of Chief Assistant at the Royal Observatory in Greenwich following H.H. Turner's election to the Savilian Chair of Astronomy at Oxford. It was in fact in Greenwich that Dyson was to spend the rest of his scientific life apart from his five years in Edinburgh. Dyson arrived at the Royal Observatory at a time when the Astronomer Royal, William (later Sir William) H.M. Christie, was busy re-equipping the observatory with new instruments. Among the smaller of these was a 13-inch photographic refractor specially designed for the photography of stars for the International Astrographic Catalogue which will be described later in connection with the Edinburgh contribution to this venture.

Large numbers of photographic plates had been collected for this undertaking before Dyson arrived in Greenwich, but the major work of dealing with these fell to Dyson's lot who in a joint paper with Christie decribed a new method by which in the computation of star positions a considerable improvement in accuracy could be achieved. In the course of this work Dyson became aware of the limitations of positional accuracy caused by the lack of knowledge of reliable stellar motions. In order to arrive at a body of well-determined stellar motions Dyson compared observations of stellar positions made at Greenwich with observations of the same stars made 80 years earlier by the English astronomer Stephen Groombridge having re-reduced the latter carefully. A catalogue of accurate values of stellar motions derived from this material turned out to be of the greatest value not only for the particular problem in hand, but for the general one of the motions of stars within the stellar system. Dyson retained his great interest in this field of work throughout his life.

Another application of Dyson's skills in positional astronomy was his reduction of the Greenwich observations of the close approach of the minor planet Eros in 1901, which A.S. (later Sir Arthur) Eddington, Dyson's successor as Chief Assistant in Greenwich, was to describe as 'a magnificent lesson in the troublous details and precautions of astronomical work of the highest precision'.

Apart from positional astronomy Dyson's other strong interest was the observation of total eclipses of the Sun for studies of the Sun's upper atmosphere. He observed three eclipses before he came to Edinburgh. His first experience was at the eclipse of May 1900 which he observed from a site in Portugal; it was the same eclipse which was observed by Copeland and his team from Spain. In the following year Dyson was in charge of an expedition to Sumatra and in 1905 he accompanied Sir William Christie on an expedition to Tunis. On all these occasions and indeed on all later eclipse expeditions Dyson's proverbial luck held: his efforts were always favoured by good weather at the critical time of the eclipse. His spectroscopic observations at these three eclipses were published in 1906 in an important paper in the *Philosophical Transactions of the Royal Society* which contains a catalogue of the positions and intensities of 1200 emission lines in the spectrum of the solar chromosphere. The paper confirmed Joseph (later Sir Joseph) N. Lockyer's earlier suggestion that the emission spectrum of the chromosphere, the Sun's upper atmosphere, is not a simple reversal of the Sun's Fraunhofer spectrum of absorption lines, but a spectrum whose lines indicate definite physical differences between the upper and lower regions of the solar atmosphere. This work brought Dyson a considerable reputation as a spectroscopist and a skilled observer of solar eclipses.

Soon after his return from the eclipse expedition to Tunisia Dyson received the Royal Warrant for his appointment to the joint post of Astronomer Royal for Scotland and Regius Professor of Astronomy in the University of Edinburgh. His Commission was for the Chair of 'Astronomy' and not any longer for that of 'Practical Astronomy', a description which had been quietly dropped by Copeland in the middle of his tenure and which was now formally abandoned.

Dyson started on his Edinburgh duties in December 1905 and moved

Sir Frank Dyson.

into his official residence on Blackford Hill with his wife, formerly Caroline Bisset Best, daughter of Palemon Best, MB, whom he had married in 1894, and their six children.

Dyson's first task on Blackford Hill apart from his supervision of the Observatory's routine work, the time service, meteorological observations and seismological recordings was the completion of the extensive programme of meridian circle observations of the positions of Zodiacal Stars which had been started by Copeland in 1898. Dyson's Annual Reports give detailed accounts of the progress of his programme which was completed in 1908 leading to a 'Catalogue of 2713 Zodiacal Stars for the Equinox 1900' which was published in 1910 as Volume III of the *Annals of the Royal Observatory*.

In 1907 the Observatory took up part of an international programme which had already been familiar to Dyson in his Greenwich days. This was the measurement of star positions on photographic plates which had been taken for the International Astrographic Catalogue or Carte du Ciel. The project of a photographic chart of the whole sky had been first raised in 1887 at an international conference in Paris on Astronomical Photography. The programme aimed at providing complete photographic coverage of the sky to stars of 14th magnitude together with a catalogue of positions of all stars down to 12th magnitude. Eighteen observatories had originally undertaken to contribute to the programme and by 1907 some like the Royal Observatory in Greenwich had actually completed the task of taking the necessary photographs in the zone of sky allotted to them and of measuring the positions of all stars on their plates. Other observatories had found the work involved in this over-ambitious undertaking more daunting than expected and had to call on assistance from outside. Amongst these was the Perth Observatory in Australia.

An agreement was reached between Dyson for the Royal Observatory Edinburgh and the Government Astronomer of Western Australia, Mr Cooke, that the Royal Observatory would take over the measurement and reduction of some of the photographic plates taken at Perth in three zones of southern declination. However, it was to be a long haul in time and expense before the results of this cooperation were to be brought finally to fruition.

Another programme started by Dyson and carried on for many years was the observation with the 15-inch refractor of double stars and in particular of stars close to the celestial pole which were inaccessible to the 28-inch refractor of the Royal Observatory in Greenwich. The results of this programme were regularly published in the *Monthly Notices of the Royal Astronomical Society*.

Dyson's interest in solar research led him to give his special support to a spectroscopic programme which had been started by Dr J. Halm and which aimed at the determination of the period of rotation of the Sun and its possible variation with heliographic latitude. When Halm started this work in 1901 he used a heliometer to throw the light of two opposite limbs of the Sun upon neighbouring parts of the slit of his spectroscope allowing him to measure double rotational Doppler displacements of lines in the solar spectrum. Dyson had Halm's visual spectroscope converted into a photographic instrument and the whole solar installation substantially improved with the result that work on the rotation of the Sun became a continuing programme of the Observatory for many years. It was carried on with more or less the same equipment long after its original investigator Dr Halm had left the Observatory to become Chief Assistant at the Cape in 1907 and it was indeed continued until the late 1930s. Results were published from time to time in the *Monthly Notices of the Royal Astronomical Society*.

It has already been mentioned that Dyson always maintained a strong interest in the motions of the stars. In 1908 he read a paper before the Royal Society of Edinburgh in which he came out in strong support of a hypothesis which had been first put forward in 1904 by the great Dutch astronomer Jacobus Cornelius Kapteyn. According to Kapteyn the motions of the stars in space are not at random in direction as had been previously assumed, but indicate the existence of two star streams moving in preferential directions. Countering the objection that apparent systematic motions could arise from observational uncertainties Dyson demonstrated by statistical methods that the existence of preferential motions could be proved from the data for stars with large proper motions where the results could not be seriously in error. Kapteyn's non-uniform pattern of motions was indeed real, and Dyson's interest was most pertinent to the elucidation of what in modern terms is described as the structure and dynamics of the Galaxy.

In 1910 Dyson published his first book, *Astronomy*, an attractive semi-popular account of the subject which was based on the lectures to undergraduates which he delivered at the University of Edinburgh. He took a great interest in teaching, and gave a course of sixty lectures and twenty meetings at the Observatory where students received practical instruction and were afterwards invited into the Dyson family circle. These visits to

Dyson's residence on Blackford Hill were the origin of the traditional summer-term tea party for astronomy students at the Observatory which still continues to this day.

In October 1910 the Observatory lost the services of its Chief Assistant, Mr Thomas Heath, who retired after 36 years' service during the whole of which time he had been absent from the Observatory for one week only and this because of illness. Mr Heath's departure marked the end of a link reaching back to the era of Piazzi Smyth in the old Calton Hill observatory. He had devoted his years to the laborious observational and computational work associated with the meridian programmes, and had served three Directors faithfully and well.

On the same date, 1st October 1910, after only five years at Edinburgh, Dyson received the Royal Warrant for his appointment as Astronomer Royal and Director of the Royal Observatory in Greenwich to succeed Sir William Christie. Dyson's short stay in Edinburgh while looked at as 'temporary exile' by his Greenwich friends, was a period full of activity and achievement during which he consolidated the status of the Royal Observatory Edinburgh on the British and international scene. He had the satisfaction of being able to complete and leave ready for printing before he left the important work on the positions of Zodiacal Stars initiated by Copeland. This was the last programme in positional astronomy carried out at Edinburgh whose meridian circle was from then on only used for time-keeping observations.

In the University Dyson equally left his mark and his departure was regretted by both colleagues and students; according to *The Scotsman* he had been one of the most popular Professors. Dyson's family also did not find it easy to leave their happy home on Blackford Hill where their seventh child had been born. In a charming biography of Dyson written in 1951 by one of his daughters, Mrs Margaret Wilson, three chapters are devoted to their Edinburgh years. However, more than twenty years of outstanding work lay ahead of Dyson in Greenwich which in Eddington's words make him 'rank among the greatest of the makers of modern astronomy'.

This is not the place to discuss Dyson's work at Greenwich in any detail. Two matters of particular general interest may be mentioned, however. The first concerns Dyson's organisation in the darkest days of the First World War of two solar eclipse expeditions for the sole purpose of testing Einstein's Relativity Theory and in particular his prediction of the deflection of light passing through the Sun's gravitational field. In 1917 in a paper to the *Monthly Notices of the Royal Astronomical Society* Dyson had drawn attention to the fact that the solar eclipse of 29th May 1919 when the Sun would be in the direction of the rich star field of the star cluster of the Hyades would provide an exceptionally favourable opportunity for testing Einstein's prediction. Dyson made all the necessary arrangements for the dispatch of expeditions from Greenwich and from Cambridge. Both expeditions were in fact able to go, the first under A. C. D. Crommelin to Brazil, the second under Eddington to Principe off the west coast of Africa. Both expeditions were successful, and at a historic joint meeting of the Royal Society and the Royal Astronomical Society on 6th November 1919 Dyson and Eddington were able to demonstrate that their results were in full agreement with

Einstein's predictions and that the tenets of his Relativity Theory had therefore been established.

A more popular example of Dyson's work in Greenwich was his concern with the Observatory's time service and his efforts to improve the accuracy of 'Greenwich Time'. The broadcasting of the six-pip time signal on BBC radio was initiated in his time and came to be known as 'Dyson's signature tune'.

Even after his retirement in 1933 Sir Frank Dyson, as he was then, remained active in more than one field of astronomy, particular evidence of this being the well-known book *Eclipses of the Sun and Moon* which he published in 1937 in collaboration with Richard (later Sir Richard) v.d.R. Woolley who in his turn was to become Astronomer Royal.

By the time he retired Sir Frank Dyson could count among the many honours which had been bestowed on him in recognition of his work, eight honorary doctorates including one from the University of Edinburgh. Having been elected a Fellow of the Royal Society in 1901, Dyson became President of the Royal Astronomical Society in 1911 and President of the International Astronomical Union in 1928. He was made a Knight Bachelor in 1915 and received the order of KBE in 1926. Lady Dyson died in 1937 and he himself died in May 1939 on his way home to England from a family visit to Australia. Like Christie, his predecessor in Greenwich, Dyson was buried at sea.

9

RALPH SAMPSON, CLOCKS, TELESCOPES AND SPECTRA

DYSON'S SHORT tenure of the Edinburgh post was followed by one of the longest in the history of the Royal Observatory, that of Ralph Allen Sampson who guided the work on Blackford Hill for 27 years.

Sampson's academic background was very similar to that of Dyson. Born on 25th June 1866, the fourth of a family of five and son of James Sampson, a metallurgical chemist, Ralph received his secondary education at the Liverpool Institute, a well-known semi-private school in Liverpool where the family had moved from Ireland. Exceptional mathematical abilities allowed young Sampson to enter Cambridge University in 1884 where he was admitted by St John's College as a sizar and elected as scholar in the following year. He graduated a year before Dyson as third wrangler in the Mathematical Tripos of 1888. In 1889 while holding a Lectureship in Mathematics at King's College London Sampson gained the first Smith's prize and in November a Fellowship of his College in Cambridge.

Having published while in London a substantial memoir on hydrodynamical theory in the *Philosophical Transactions* Sampson returned to Cambridge in 1890 where he gained one further distinction in becoming the first holder of the newly established Isaac Newton Studentship in Astronomy and Physical Optics which a year later was to be bestowed on Dyson. In Cambridge Sampson made his first excursion into astrophysics and in particular into astronomical spectroscopy with the help of H. F. Newall, who was then Senior Demonstrator in the Cavendish Laboratory under Professor J. J. Thomson and who was later in 1913 to become the first Professor of Astrophysics in the University of Cambridge.

Sampson was first and foremost a mathematician and the principal result of his Cambridge work was a theoretical paper 'On the Rotation and Mechanical State of the Sun' which appeared in 1895 in the *Memoirs of the Royal Astronomical Society*. It was a memorable publication because it introduced for the first time the theory of radiative equilibrium into astrophysics showing that much of the heat generated in the deep interior of the Sun is carried outwards not by streams of moving gas but by radiation, even at depths where the gases are highly opaque.

In the autumn of 1893 while Dyson was preparing to leave Cambridge for the Royal Observatory in Greenwich Sampson was elected into the Chair of Mathematics in the Durham College of Science at Newcastle upon Tyne. Two years later, on the retirement of Professor R. J. Pearce, Sampson moved to Durham itself as Professor of Mathematics and Director of the Durham Observatory, a small institution which, however, had then been in operation for some fifty years. Sampson's interest in this observatory led to his mounting there of what became known as the Durham Almucantar, an instrument in which meridian transits of stars were observed across a horizontal circle instead of a vertical wire in the meridian. The ingenious instrument which made it possible to do away with some of the instrumental errors of normal transit circles, attracted considerable attention at the time and for some years was used for observations of the variation of latitude.

In his undergraduate days Sampson had been a pupil of John Couch Adams, the Lowndean Professor of Astronomy in Cambridge who in 1846 had shared with Le Verrier in Paris the honours of the great discovery of the planet Neptune from his theoretical investigations of irregularities in the motion of the planet Uranus. John Couch Adams was one of the leaders in the field of dynamical astronomy, and it was to this field that Sampson made one of his most outstanding contributions which in due course was to lead to the award to him of the Gold Medal of the Royal Astronomical Society. This work which he started in Durham was concerned with the interpretation of the motions of the four great satellites of Jupiter.

The movements round Jupiter of its four great satellites which had first been discovered in 1610 by Galileo Galilei, had been investigated by many of the great dynamical astronomers of the 18th and 19th centuries. This was because with their relatively short periods of revolution round Jupiter – 1.8 days for the nearest satellite Io and 16.7 days for the most distant of the four, Callisto – as compared with the periods of revolution of the planets round the Sun the Jupiter system offers considerable advantages over the solar system when one wishes to study the effects on orbits of long-lasting perturbations. There is also a curious relation between the periods of revolution round Jupiter of its three inner satellites which was first discovered by the great French astronomer Laplace.

Top: Objective prism photograph taken on 20 March 1912 of the spectra of Nova Geminorum and the neighbouring A2-type star Theta Geminorum. *Bottom:* Comet Brooks photographed with the 15-inch refractor on 27 September 1911.

Sampson became interested in the problem of Jupiter's satellites a few years after his arrival in Durham by which time serious discrepancies between theoretical predictions and actual observations of the orbits of the four satellites had become obvious. The disagreements could be due to inadequacies in the theory or imperfections in the observations which at the time consisted of visual recordings of the times of the eclipses of the satellites by the disc of Jupiter. Having been supplied by Professor E.C. Pickering of the Harvard Observatory with a series of more accurate photometric observations of satellite eclipses Sampson used these to amend the existing theory of satellite orbits. In spite of these improvements there remained, however, substantial differences between theory and observation which according to Sampson could only be dealt with by working out a new dynamical theory. This major task was completed by him in 1910 when the *Tables of the Four Great Satellites of Jupiter* which were based on the new theory and contain on 350 pages the positions of the satellites from the year 1850 to the year 2000 were published by the University of Durham. His 'Theory of the Four Great Satellites of Jupiter' appeared only in 1921 in the *Memoirs of the Royal Astronomical Society*.

A task difficult in other ways which occupied much of Sampson's time in Durham was the editing of the unpublished manuscripts of his old tutor, John Couch Adams, which were to be brought out by the Cambridge University Press. These were in fact published by Sampson as the first part of the second volume of *Adams' Collected Works*. Sampson also saw to it that Adams' important lectures on the 'Lunar Theory' and his manuscripts on the 'Perturbations of Uranus' leading to the discovery of Neptune were published separately. The considerable success of Sampson's editorial work led the Cambridge University Press to offer him the post of Editor-in-Chief for their project of a complete edition of the scientific publications and correspondence of Isaac Newton, an offer which he declined only on account of his involvement at the time in the heavy work on Jupiter's satellites.

While he was still in Durham Sampson's achievements were acknowledged by his election into the Royal Society in 1903 and by the conferment in the following year of the honorary Doctorate of Science on him by the University of Durham which in 1908 also revived in his favour the office of Professor of Astronomy which had been dormant since the retirement in 1871 of Professor Chevallier.

By 1910 Sampson's scientific reputation was such that no one was surprised when in December of that year he was appointed to succeed Frank Dyson in the joint post at Edinburgh. He was the first to come to the Professorial part of it from a Chair at another University, but having done his most important work in the field of celestial mechanics, was he perhaps the man who would change the work of the Royal Observatory in the direction of classical astronomy? As matters turned out the very opposite happened. Under his guidance the Observatory moved into the very newest fields of astronomy and there can be little doubt that the 27 years which Sampson spent on Blackford Hill were amongst the most successful in the history of the Royal Observatory.

Summaries of the Observatory's work under Sampson can be found in

R. A. Sampson.

the 27 Annual Reports which he submitted to the Secretary of State for Scotland between 1911 and 1937. Leaving aside the continuation and completion of earlier programmes such as the observation of double stars, the spectroscopic determination of the Sun's rotation or the measurement of the plates for the Perth section of the *Astrographic Catalogue* Sampson's interests centred on three problems: the improvement of the determination of Time, the progress in the optical performance of telescopes, and the introduction of objective methods into the photometry and quite particularly into the spectrophotometry of stars.

The determination of time by the observation with transit circles of the passages of stars across the Meridian due to the Earth's rotation had been one of the fundamental regular observations made at nearly every observatory in the world from at least the beginning of the 19th century. Pendulum clocks whose errors were determined by such astronomical observations were used to provide a regular time service of one sort or another which could include as in Edinburgh the firing of a time-gun and the dropping of a time ball.

When Sampson took over the direction of the Royal Observatory its daily time-service for the cities of Edinburgh and Dundee was based on the use of a single pendulum clock made by Molyneux which supplied time to clocks at the General Post Office, at Edinburgh Castle and at Nelson's Monument on Calton Hill. Failures in the service led Sampson to the acquisition of a standard clock of high precision constructed by Dr Sigmund Riefler of Munich in Germany. The acquisition of this clock was made the occasion for a careful revision of everything connected with the recording of time. As

Sampson put it in one of his Annual Reports: 'The Clock is one of the fundamental measuring instruments of an observatory and deserves a much closer study than is usually given to it'. He introduced a system by which the behaviour of the various clocks in the observatory was regularly monitored and their rates daily compared by means of a microchronograph to an accuracy of a thousandth of a second. The temperature control in the clock chamber was considerably improved and a comparison of the Edinburgh clocks with standard clocks in other institutions was made possible by the installation in 1913 of equipment for receiving wireless time signals. Sampson's interest in clocks led to several substantial papers of which the first on 'Studies in Clocks and Time Keeping' appeared in 1917 in the *Proceedings of the Royal Society of Edinburgh*, and the second in 1918 'On the Measurement of Time to a Thousandth of a Second' in the *Monthly Notices of the Royal Astronomical Society*. These papers were followed in 1924 by a Memoir which Sampson communicated to the Royal Society of Edinburgh and which contained the results of all the more important comparisons between four standard clocks of the observatory and observations of time with the observatory's transit circle.

Among the clocks was one constructed by a civil engineer, W. H. Shortt, in association with the Synchronome Company which had been placed in the Royal Observatory for trials in 1923. This Free Pendulum Clock, keeping time to an accuracy hitherto unknown turned out to be superior to all others. It had two precisely synchronised pendula, a 'Master' which had to do no work and was kept swinging in a vacuum chamber at constant temperature by tiny impulses imparted to it at intervals, and a 'Slave' which looked after the turning of the counting wheel and the operation of the dials. It was as a result of Sampson's tests of this clock that Shortt Free Pendulum Clocks became and remained for a long time the standard time keepers in many observatories. The free pendulum clock was the most accurate clock ever devised, the first clock to detect small irregularities in the rotation of the Earth which up till then had been considered the most perfect possible time-keeper. The original clock known as Shortt No.0 was presented to the Royal Observatory by the Synchronome Company. Together with Shortt No.4 it was regularly used in the Observatory until the time that pendulum time keepers had to give way to quartz clocks when the Edinburgh Shortt clocks were presented to the Royal Scottish Museum in 1963.

Sampson's fundamental work on the precision of time determinations which he pursued vigorously in the difficult conditions of the First World War was acknowledged by his election as the first President of the 'Commission de l'Heure', the international committee founded to study the problem of astronomical time-keeping.

While still in Durham Sampson had already become interested in theoretical optics and its application to the improvement of telescopes. In Edinburgh this interest was resumed when he tried to bring the old 24-inch reflector in the western tower into regular use and found that the figure of its glass mirror was poor, but that its thinness did not allow correction by repolishing. With the view to replacing this telescope by a new reflector with the best possible optical design Sampson embarked on fresh studies of optical aberrations resulting in two substantial Memoirs which were pub-

Left: The Molyneux mean time clock which provided the local time service from Calton Hill and Blackford Hill for over half a century (Royal Scottish Museum).
Right: The Shortt free pendulum clock No.0 master and slave, installed at the Royal Observatory Edinburgh in 1923 (Royal Scottish Museum).

lished in 1913 and 1914 in the *Philosophical Transactions of the Royal Society*. In the second of these he suggested an optical arrangement for a Cassegrain type of reflector in which the interposition of a pair of suitable lenses in the outcoming beam of light could reduce all optical aberrations to a minimum. A further application of the same principles to the correction of the field of a Newtonian reflector was published by Sampson at the same time in the *Monthly Notices of the Royal Astronomical Society*.

The outbreak of the First World War and the difficult conditions in the early post-war years made it impossible for Sampson to see his plans for a new telescope become reality. This had to wait until 1927 when his proposals for an extensive re-equipment of the Royal Observatory were accepted by HM Treasury.

In a remarkable effort to make some use of the 24-inch reflector in a programme in which the faulty figure of the mirror and the resulting poor image quality might not matter very greatly, Sampson decided to embark on photoelectric stellar photometry. At the time in 1915 this was an entirely new field of astrophysics which had been pioneered only a year or two before by P. Guthnick in Berlin-Babelsberg and J. Stebbins at the Washburn Observatory, Madison. Its pursuit required a considerable amount of experimentation in the laboratory which was carried out by Dr E. A. Baker, then Junior Assistant, who constructed and tested a number of sensitive potassium, sodium and later caesium and rubidium photoelectric cells of the type which had been invented in 1911 by Elster and Geitel in Germany. For the measurement of the feeble photoelectric current Sampson like Guthnick used a sensitive filar electrometer.

The aim of this first photometric programme of the Royal Observatory was changed by Sampson in 1920 when the direct photoelectric determination of stellar brightness at the telescope was replaced by the photoelectric measurement of the photographic densities of the images of stars on photographs by means of a microphotometer in which a beam of light passes through the photographic plate before it falls on a photoelectric cell.

In 1921 one further and final change was made by Sampson in the photometric work of the observatory. His aim was to measure the photographic density not of the images of stars but of the images of spectra by passing a beam of light through spectrograms and recording the effect on tracings. Through Sampson the recording microphotometer became the standard instrument in stellar spectrophotometry and the work which he started in this field became in the course of time and with ever improved equipment the principal programme of work in the Royal Observatory which continued for some forty years.

Sampson's purpose was the determination of the spectral distribution of intensity in different types of stars measured in terms of that of the standard star Polaris whose spectrum was photographed first and last on each plate. His early spectrograms were obtained with a 12 degree objective prism which he mounted in front of a 6-inch photovisual refractor. The modest equipment allowed only spectra to be obtained of stars brighter than the third magnitude, but a substantial number of spectrograms with an average of 10 spectra for each star was obtained in the course of time and the results of this pioneering work were published by Sampson in a preliminary paper in 1923 and two major papers in 1925 and 1930 in the *Monthly Notices of the Royal Astronomical Society*. The stellar observations were supplemented by an important laboratory investigation by Dr Baker who studied the precise relationship between photographic density and light intensity under a variety of different conditions. The results of this work appeared between 1925 and 1928 in three papers in the *Proceedings of the Royal Society of Edinburgh*. The whole spectrophotometric programme led to the determination of temperatures of stars ranging in colour from blue to red in a range of spectral types from Bo to Mo.

This remarkable programme was terminated only in 1929 when the expected major re-equipment of the Observatory was about to take place.

The 36-inch Cassegrain telescope installed in the East dome in 1928.

As the most important part of it Sampson had proposed a new reflector and a universal slit spectrograph. He succeeded in obtaining a 36-inch reflector constructed by Sir Howard Grubb Parsons and a spectrograph which could be used with one, two or three glass or quartz prisms and with interchangeable long, medium and short focus cameras. The new 36-inch telescope was mounted in the East Dome of the Observatory in 1930 where it replaced the 15-inch refractor which was re-erected in 1932 on a new mounting and with a 10-inch, f/6, camera attached to it. The 36-inch telescope was completed by the delivery in March 1931 of the new spectrograph which had been constructed by Messrs A. Hilger Ltd.

Sampson's last years in the Royal Observatory were occupied in seeing the new equipment brought into full operation and used for an extension to fainter stars of his earlier spectrophotometric work. In 1934 he had the satisfaction of being able to use the 36-inch telescope and spectrograph for an exceptionally good set of spectra of Nova Herculis which were published by Professor F. J. M. Stratton of Cambridge as part of his collection of spectrograms of the Nova. One of Sampson's last efforts in the Observatory was concerned with the construction of an ingenious Blink-Comparator to a design by Dr Baker which was built in the Observatory workshop and used for the detection of variable stars on plates taken with the 10-inch camera.

Those who knew Sampson have made a point of saying that one of his most striking characteristics was his capacity for work. While he was kept busy at the Observatory he was equally concerned with his Professorial duties and with various activities outside. He devoted much time and effort to a comprehensive course on astronomy in the University where a philosophical turn of mind which showed itself frequently in his later years came out already in his Inaugural Lecture which he delivered in 1911 on 'The Growth of Ideas in Astronomy'. With a keen interest in the history of science he took a very active part in the Napier Tercentenary Celebrations in 1914. Being a member of Council for some twenty years including the years as General Secretary Sampson was much involved in the affairs of the Royal Society of Edinburgh which awarded him its Keith Prize. In London he guided the Royal Astronomical Society as President from 1915–17.

Failing health forced Sampson to retire in September 1937 when he was in his 72nd year. He was one of the last Professors in the University of Edinburgh to have been appointed for life – 'aut vitam aut culpam', as the formula went. He and his wife Ida whom he had married in 1894 left their residence on Blackford Hill where they had raised a family of five children first to travel abroad and then to settle in Bath where he died on 7th November 1939. One of his daughters, Professor Peggy Sampson, was to become a celebrated musician and a well-known player of the Cello and the Viola da Gamba.

Sampson was a great scholar and a remarkable pioneer whose outstanding researches in both classical astronomy and modern astrophysics gained for the Royal Observatory the highest reputation. The equipment he installed was the mainstay of the Observatory for decades and his initiation of photographic photometry and spectrophotometry moulded the course of the Observatory's work making it a leading centre in this field.

10

W.M.H. GREAVES AND MORE SPECTROPHOTOMETRY

AFTER A brief interregnum in which Mr J. Storey, First Assistant, was in charge of the Observatory Professor Sampson was succeeded on 1st April 1938 by William Michael Herbert Greaves who like Dyson came from the Royal Observatory in Greenwich and who like Dyson and Sampson had received his academic training at the University of Cambridge. Greaves' appointment was widely expected at the time since in Greenwich he had been in charge of a programme on the determination of stellar temperatures which fitted well into the Edinburgh scene.

Greaves was born in Barbados, British West Indies, on 10th September 1897, the only son of Dr E. Greaves, a medical graduate of Edinburgh University. Having spent his school years at Barbados concentrating on the study of Mathematics and English Greaves won the Barbados English Scholarship in Mathematics with which he came to England and entered like Sampson St John's College, Cambridge, in 1916 to read Mathematics. His mathematical prowess showed itself when in his Finals three years later he became a Wrangler with Distinction in Part II of the Mathematical Tripos. At the same time he was awarded the Tyson Gold Medal for excellence in astronomical and allied subjects and in 1921 he gained like Dyson and Sampson before him both the Smith's Prize and the Isaac Newton Studentship. He became a Fellow of his College in the following year.

In his research at St John's Greaves came under the influence of the distinguished mathematician H.F.Baker, who was then Lowndean Professor of Astronomy and Geometry. He directed Greaves' early researches into the field of classical celestial mechanics in which Greaves published in 1922 as many as five papers, four in the *Monthly Notices of the Royal Astronomical Society* and one in the *Proceedings of the Cambridge Philosophical Society*. In the first of these he dealt afresh with an old problem which had first been investigated in the late 18th century by the great French mathematician Joseph Lagrange. It concerned the orbits of asteroids which move in the same plane as Jupiter and whose distances from the Sun are nearly equal to their distances from the Planet so that they form an equilateral triangle with Sun and Jupiter. In his other papers Greaves discussed perturbations of orbits caused by motions which are commensurable such as the effect of the planet Mars on the orbits of asteroids which move round the Sun at half the speed of Mars, or the effect of Mimas and other satellites of Saturn on the orbits of those particle constituents of Saturn's rings which move round Saturn at half the speed of Mimas. This, of course, is a problem which has become very fashionable once again as the result of recent space probe observations of the complex structure of Saturn's rings.

At St John's College in Cambridge Greaves came first into contact with Edward (later Sir Edward) V.Appleton who at that time was Demonstrator in the Cavendish Laboratory where he studied the problem of generating radio waves with the help of triode valves. The theory of the triode oscillator implied the mathematical solution of a set of differential equations which were similar to those which Greaves had tackled in his work on orbits of the particles of Saturn's rings. Greaves' natural interest in the triode oscillator resulted in several papers, all published in 1923, the first jointly with Appleton in the *Philosophical Magazine*, the others in the *Proceedings of the Royal Society* and the *Cambridge Philosophical Society*.

While Greaves was still in Cambridge Professor Sir Joseph Larmor of St John's College introduced a scheme by which Isaac Newton students could spend some of their time at the Royal Observatory in Greenwich where they could be introduced into the practical work of astronomers. Greaves was one of the two beneficiaries of the scheme – the other being L.J.Comrie, later Superintendent of the *Nautical Almanac* and brother-in-law of Greaves – and was fortunate enough to become closely associated with an outstanding Greenwich astronomer, C.R.Davidson, a Fellow of the Royal Society. Greaves' work as Isaac Newton student in Greenwich led already early in 1924 to his appointment to the post of second Chief Assistant in succession to Harold (later Sir Harold) Spencer Jones who had become HM Astronomer at the Cape. For the next nine years he worked side by side with the first Chief Assistant, John Jackson, succeeding him when the latter took over the direction of the Cape Observatory on the return to Greenwich of Spencer Jones as Astronomer Royal.

Greaves spent fourteen years at Greenwich during which he became concerned with a variety of problems amongst which stellar spectrophotometry and the determination of temperatures of stars – like Sampson's earlier work at Edinburgh – grew into his very special interest.

Of other investigations carried out by Greaves in Greenwich mention should be made of his joint studies with H. W. Newton of the effects of sunspots on terrestrial magnetism. Their work was an extension of earlier Greenwich investigations by E. W. Maunder who in 1904 had found a close correlation between the number of sunspots visible on the solar disc and the occurrence of major magnetic disturbances, magnetic storms, on the Earth. On the basis of substantially greater observational material Greaves and Newton were able to confirm in three papers published in 1928 and 1929 in the *Monthly Notices of the Royal Astronomical Society* two of Maunder's principal results: that great magnetic storms are generally connected with the appearance of large sunspots and commence suddenly about 30 hours after the sunspot has moved across the central meridian of the Sun; and that magnetic storms have a tendency to recur at intervals of 27 days when the Sun's rotation has brought the same solar region back into view. Greaves and Newton could show, however, that the 27-day recurrence does not apply to great storms. Greaves kept an interest in solar-terrestrial relationships for many years long after he had left the Royal Observatory.

It has already been mentioned that Greaves' main interest in Greenwich which he kept up all his life was the investigation of the problem of stellar temperatures. When he started this work in 1925 the solution of the problem though troublesome in practice, appeared relatively simple in theory because as Greaves himself put it thirty years later 'the whole subject was dominated by the belief that the radiation emitted by a star was distributed in wavelength roughly according to Planck's black-body formula'. If this were actually the case a star's temperature could be derived from a measurement of the ratio of the intensities of radiation emitted at two different wavelengths in its spectrum. However, it has been known for a long time that the problem is considerably more complex and that such measurements lead to the derivation of 'colour temperatures' only, whose values depend on the spectral regions of their measurement.

Greaves working together with C. R. Davidson and later also E. G. Martin chose for their investigation two regions in the red and blue parts of the continuous spectrum which were as free as possible of absorption lines. The team measured the spectral intensities in those regions for different types of stars comparing them with each other and in order to obtain absolute values of temperatures comparing also the intensity distribution in a set of 'standard stars' with that of a terrestrial light source of known temperature. The technical difficulties in this programme were very considerable. Since the different spectra were obtained by photography the vagaries of the photographic plate in its response to radiation had to be studied and allowed for carefully. A major problem was posed by the effect on observations of absorption in the atmosphere which in the situation of Greenwich close to London was found to vary considerably. Another problem arose in the

W. M. H. Greaves

comparison of the stellar spectra with the spectrum of the terrestrial source which the team had mounted on the roof of the old octagon room of the Royal Observatory. This was about 600 feet away from their telescope for which they used first the Thompson 30-inch reflector and later the new 36-inch Yapp telescope which had been presented to the Observatory in honour of Sir Frank Dyson.

The Greenwich programme on colour temperatures of stars carried through by Greaves and his team was at the time the most substantial of its kind and was rivalled only by Professor H. Kienle's work at the Göttingen Observatory in Germany. The Greenwich observations led to a number of papers in the *Monthly Notices of the Royal Astronomical Society* and also in *The Observatory*. The results were summed up in two substantial volumes of *Greenwich Colour Temperature Observations* of which the first appeared in 1932 and the second much delayed by the war in 1952 after Greaves himself had left Greenwich. By-products of this spectrophotometric programme were papers by Greaves and Martin on the spectra of individual stars and on the problem of the so-called yellow B-stars. At that time the presence of interstellar dust as the cause of reddening or yellowing of distant intrinsically blue stars was still under discussion.

When Greaves arrived in Edinburgh in 1938 and had to consider the future activities of the Observatory he decided that for the time being at least most of the routine work and programmes like the search for variable stars in Kapteyn's Selected Areas of the sky should be continued and that the Edinburgh work on the Perth Astrographic Plates should be prepared for the press and published with a grant which he obtained from the International Astronomical Union.

At the same time Greaves was anxious to go ahead with a new spectrophotometric programme which was to be devoted not to the photometry of continuous spectra and the determination of stellar temperatures, but to the measurement of the intensities and profiles of absorption and emission lines and their dependence on physical conditions in stellar atmospheres. The programme made a very promising start, but when the Second World War broke out in 1939 and a general black-out had to be declared the Observatory's observational work had to be slowed up and ultimately suspended altogether. This happened to the spectrophotometric work at the end of 1940 and two years later also to the programme on variable stars in Kapteyn's Selected Areas. However, some first results of the latter programme were ready for publication in the *Monthly Notices of the Royal Astronomical Society* in the late 1930s and a description of the instrumentation used for the programme, including that of a specially designed blink-comparator, appeared in 1940 in the first volume of the new *Publications of the Royal Observatory Edinburgh*.

By 1943 the work of the Royal Observatory was reduced to routine activities such as the local time service and the maintenance of meteorological and seismological records. However, a new and important programme of work had appeared for the observatory in the meantime. In the autumn of 1940 when the country became the target of severe air raids Sir Harold Spencer Jones, the Astronomer Royal, suggested that the Edinburgh Observatory ought to be equipped for the provision of an independent national time-service in case the normal service from Greenwich's station at Abinger were put out of action. Greaves took up this major new task with enthusiasm and managed with the sole additional assistance of two members of the Greenwich staff to complete the installation of all the necessary equipment by the end of the year and to operate the Rugby rhythmic time signals in January 1941 when the Greenwich time-service was in fact disrupted by enemy action.

Greaves kept this national time service going throughout the whole war checking the standard clocks whenever possible by transit observations. As a by-product of this work he made a careful comparison of the performance of the Edinburgh and Greenwich free pendulum Shortt clocks with that of quartz crystal clocks newly installed by the Post Office. The result, showing the definite superiority of quartz over even the very best pendulum clocks was published by Greaves and Symms, one of the Greenwich staff, in 1943 in the *Monthly Notices of the Royal Astronomical Society*. The same journal carried in 1945 and 1946 Greaves' detailed accounts of Edinburgh's contribution to the war-time national time-service which was suspended only in February 1946.

After the end of the war the research work of the Royal Observatory and in particular its spectrophotometric programme was resumed. The programme covered about a hundred blue stars of O- and B-types down to 5th magnitude. Their spectra were photographed with the new equipment which Sampson had acquired, the 36-inch reflector in conjunction with the Hilger spectrograph in its two-prism arrangement. The spectra were photometrically calibrated with the help of a separate multiple-slit spectrograph and a laboratory source.

For the reduction of the observational data Greaves with the help of Dr E. A. Baker introduced a novel method of measurement which was to

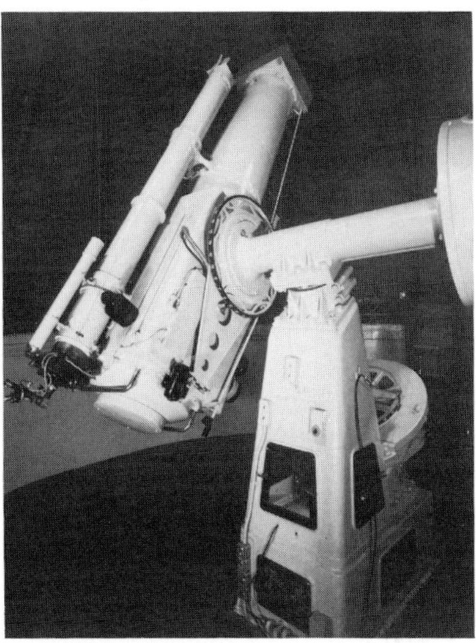

Left: The small transit telescope used for a back-up British time service during the years of the Second World War. *Right:* The 16/24-inch Schmidt telescope installed in 1951. The pier was erected in 1920 for the 10-inch astrograph.

reduce effects of plate grain and random and systematic errors introduced by the conventional method of smoothing microphotometer tracings. The stellar spectrograms were measured in Greaves' programme at a number of fixed wavelengths, up to 1500, the same for all spectra of a given type. These measures were then superimposed numerically – stacked in present-day language – first for the several spectrograms of each star, and secondly for the spectrograms of all stars of a given type. The measurements were made with a manually operated microphotometer specially designed and made in the observatory workshop. It was clear that such a method could lead to appreciably greater accuracy than conventional procedures and was capable of discovering very faint spectral features, but at the same time it was bound to give rise before the advent of computers to extremely laborious and time-consuming work.

This spectrophotometric programme on line intensities in stellar spectra remained in the foreground of observatory activities for many years and was ultimately to result in seven major observatory publications. The first paper of the series was published after the end of the war in 1949 when Baker described in the second number of the first volume of the new *Publications of the Royal Observatory Edinburgh* the specialised techniques of measurement and reduction of spectrograms which Greaves had introduced and which were never changed throughout the whole programme. That first paper contained, apart from results on the intensities of lines in the red and blue regions of the spectra of hot Oe5-type stars, an important account of the emergence of the 'Delta Function', which is a density transform providing a nearly linear relation between photographic density and the intensity of the radiation causing it. The same paper contained the very first precise data on the broad absorption feature which is observed at a wavelength of 4430Å and which is caused by material in interstellar space, an example of the advantage of Greaves' methods over conventional ones.

The second paper of the series by Greaves, Baker and Wilson appeared in 1955; this was in fact Greaves' last scientific publication. Two further papers followed by Dr R. Wilson (now Professor of Astronomy at University College London) who had joined the staff of the Observatory in 1949. Dr Baker retired in 1953 after forty years of service and was replaced by Dr H. E. Butler who looked after the completion of Greaves' spectrophotometric programme which involved the measurement and reduction of some 800 spectrograms of a hundred blue stars ranging in spectral type from Oe5 to B7.

After the war Greaves also decided, harking back to his earlier interest in the field of solar-terrestrial relations, to launch a study of temporary solar phenomena such as flares and active prominences. In October 1947 he appointed in succession to Mr R. W. Wrigley who had completed the Perth Catalogue work, Dr M. A. Ellison as Principal Scientific Officer to be in charge of the new solar work. Before joining the staff of the Royal Observatory Dr Ellison had been a Science Master at Sherbourne School where he had devoted all his spare time to solar studies using a spectrohelioscope of his own construction. This instrument had been designed for visual or photographic observations of short-lived solar phenomena such as flares in sunspot regions which could be studied in the light of the red hydrogen H-Alpha line. It was first set up in 1947 by Dr Ellison in the optical room of the observatory where it was fed with sunlight by a siderostat mounted outside on the roof. Five years later the installation with improved optics and a new coelostat was moved to a better site in the now disused Transit House of the observatory.

Regular observations of the Sun with Ellison's spectrohelioscope were started in the spring of 1948 and in the following few years visual observations as well as photometrically calibrated plates of spectra of many flares and prominences were obtained. To study correlations between solar activity and terrestrial effects and in particular between solar flares and sudden enhancements of atmospherics a long-wave receiver was installed at the Observatory by the Cavendish Laboratory, Cambridge, which, recording automatically, provided a continuous record of the effects of the ultraviolet radiation of flares on the state of the Earth's upper atmosphere. The results of this work were published by Dr Ellison in 1950 in the first volume of the *Publications of the Royal Observatory*.

This paper was followed in 1952 by a major photometric survey of solar flares, plages and prominences which Dr Ellison published in the same volume. This publication contains a detailed account of the visual and photographic procedures used by Ellison in the observation of flares and active regions in the neighbourhood of sunspots and includes the spectrophotometry of solar prominences at the limb and on the solar disc which were carried out with the Edinburgh spectrohelioscope by Mary T. Conway (now Mrs Brück). The prominence work was continued by Dr Ellison until he left the Observatory in 1958. In this period also he published his book

The Sun and its Influence.

An unusual phenomenon occurred on 26th September 1950 when a blue Sun was observed from Blackford Hill – a phenomenon incidentally which had engaged Piazzi Smyth's attention in Palermo. Studying the distribution of intensity in the spectrum of the blue Sun Dr Wilson was able to conclude that the observed extinction of part of the Sun's light could be explained as the result of the combination of strong neutral absorption by carbon particles and scattering by tiny globules of oil. The physical cause of the blueing was in fact a smoke layer at an altitude of some 30,000 feet whose source lay in extensive Canadian forest fires burning in Alberta.

Another interesting short programme carried out in the 1950s was the photoelectric recording of stellar scintillation – the twinkling of the stars – with the object of studying the phenomenon on a quantitative basis. Records of the scintillation of the images of stars and planets were obtained at the focus of the 36-inch reflector under different atmospheric conditions. The results of both these investigations were published in the *Monthly Notices of the Royal Astronomical Society*.

The observational facilities of the Observatory were significantly enlarged in 1951 by the acquisition of a 16/24-inch Schmidt camera which Greaves had ordered from Messrs Cox, Hargreaves and Thomson as a replacement for the 10-inch astrograph in the west dome. This was one of the first Schmidt telescopes to be introduced into a British observatory and an instrument which in later years was to play a major part in the Observatory's work. In the first few years after its installation tests and improvements occupied so much time that Greaves himself, unfortunately, was unable to witness the successful use of the new telescope.

In his later days Greaves who in 1943 had been elected into the Royal Society became very much an elder statesman whose advice was sought in the Councils of the Royal Society of Edinburgh where he served as Secretary and Vice-President and, quite particularly, in the activities of the Royal Astronomical Society which he had joined as a young man in 1921, whose meetings in London he never missed and of which he became President in 1947 after many years as member of Council, Secretary and Vice-President. On the international scene Greaves was a prominent member of the International Astronomical Union whose Assemblies he attended regularly. Photometry and spectrophotometry were, of course, the fields in which Greaves held undisputed mastery. His 1945 Halley Lecture on 'Photometry as a Weapon of Astronomical Research' and his 1948 Presidential Address to the Royal Astronomical Society on 'The Photometry of the Continuous Spectrum' remained fundamental treatises on the subject.

When the post-war needs of British astronomy came up for discussion Greaves was one of the few astronomers who came out in full support for the then new field of radio astronomy. In fact, he was a strong advocate of Bernard (now Sir Bernard) Lovell's ambitious proposal for the construction of a large steerable radio telescope at Jodrell Bank when this project came before a meeting in February 1950 in the rooms of Greaves' friend, Sir Edward Appleton, Principal and Vice-Chancellor of the University of Edinburgh. In later years Greaves was to represent the Royal Astronomical Society on the Steering Committee of the Jodrell Bank Experimental Station

Sir Edward Appleton, Principal of Edinburgh University from 1948 to 1965.

and to become as valued a member of that Committee as he had been on the Board of Visitors of the Royal Observatory in Greenwich. Greaves enjoyed being on the closest possible terms both personally and professionally with his contemporaries, yet at the same time maintaining an independent course for the Royal Observatory Edinburgh.

Greaves very much enjoyed his Professorial duties. The main astronomy course while addressed chiefly to first-year students was anything but elementary as aspirants to this class were required to have attained a certain level in mathematics and indeed the course attracted many mathematicians. Greaves also introduced an advanced course in Theoretical Astrophysics for final-year Physics and Mathematical Physics students. Lectures were given in the old Mathematics Department in Chambers Street where Greaves felt very much at home among his friends who included Professors Sir Edmund Whittaker and Alexander Aitken and Dr (now Professor Emeritus) W. L. Edge. In 1950 he had the satisfaction of seeing the appointment of the first full-time Lecturer in Astronomy, Dr M. J. Smyth, a Cambridge graduate, later to become Senior Lecturer in the Department.

In August 1955 Greaves took a prominent part in the proceedings of the General Assembly of the International Astronomical Union in Dublin where he presided over the Commission on Stellar Photometry. Nobody could have foreseen that this was to be his last contribution to the subject nearest to his heart and which he had enriched so greatly. On Christmas Eve 1955 he died suddenly and quite unexpectedly at his home on Blackford Hill at the early age of 58. His wife, formerly Caroline Grace Kitto, sister of the distinguished Greek scholar, the late H. D. F. Kitto, was well known for

Transit circle constructed by Messrs Troughton and Simms for Lord Crawford in 1873 and moved from Dunecht to Blackford Hill in 1896. This circle, with an object glass of 8.5 inches in diameter, was used for various programmes including the observations for the Edinburgh Catalogue of Zodiacal Stars.

her friendship to Greaves' many colleagues and students. She remained in Edinburgh until her death in 1978. Their only son, Dr George Greaves, a mathematician like his father, after a brilliant career at Edinburgh University and in his father's old College, St John's in Cambridge, is now a Lecturer in Mathematics at University College, Cardiff.

I had the privilege of succeeding Professor Greaves in September 1957. In the intervening period during which Dr Ellison was Acting Director the two main programmes of the Observatory, the stellar spectrophotometry and the solar work, continued without interruption as did the University teaching. The scientific staff at this time consisted of Drs Ellison and Butler as Principal Scientific Officers, Dr Wilson and Mr Seddon as Scientific Officers, and Dr V.C. Reddish as University Lecturer in Astronomy.

Part Three

The Time of New Technology and Change
1957–1975

11

NEW INSTRUMENTATION AND EXPANSION

AT THE beginning of February 1957 I received the Royal Warrant appointing me to succeed Professor Greaves in the joint post of Astronomer Royal for Scotland and Regius Professor of Astronomy in the University of Edinburgh and on 8th September of that year I took up my new appointment. When the Royal Warrant reached me I was in charge of the Dunsink Observatory in Ireland, an old 18th-century institution of Trinity College Dublin, which I had been asked to re-establish as part of the new Dublin Institute for Advanced Studies.

I looked forward enthusiastically to the opportunity of serving both the Royal Observatory and the renowned University of Edinburgh and to follow in the footsteps of my old friend Professor Greaves whose work in astronomical photometry I had long admired. I also could hope to assist my British colleagues in their efforts to rejuvenate optical astronomy in the country where, it was generally agreed, it had through lack of observational resources fallen far behind the brilliant work of radio astronomers and theoretical astrophysicists. At a very early stage Sir Edward Appleton, at the time the Principal of the University, brought up the possibility of starting at the Royal Observatory work in radio astronomy, a subject in which he himself was deeply interested, but I took the view that this field was already very well looked after by the flourishing radio establishments in Cambridge and in Manchester.

I was fortunate in that my actual arrival in Edinburgh happened to coincide with the launching, on 4th October 1957, of the first artificial Earth satellite, the Soviet Sputnik. This was an event which not only stimulated popular interest in astronomy – thereby making it easier to secure financial support for astronomical research – but also drew general attention to the remarkable advances which had been made in the war and post-war years in the field of technology.

When I came to Scotland particularly striking progress was being made in electronic engineering and I decided that at the Royal Observatory we ought to benefit from that progress and start systematic work on the design and construction of new instrumentation with the help of professional engineers qualified in the new technology. The particular aim was to introduce methods of automatic measurement and data processing into the various fields of astronomy and astrophysics where they were needed.

The execution of this instrumentation programme at the Royal Observatory – which in its first years did not find universal favour within the British astronomical community – led in the course of years to a major expansion of the Observatory staff, from less than ten members at my arrival to well over a hundred when I retired 18 years later. It also involved the erection of various new buildings for laboratories and workshops, the first extensions to the original Observatory building in sixty years.

Before any new work could be started after my arrival in Edinburgh it was necessary to complete the long-standing spectrophotometric programme of the Observatory on line intensities in the spectra of early-type stars which Professor Greaves had initiated on his arrival in 1938 and which had taken up much of the Observatory's time in the intervening years. By 1957 four major papers had been published on this work by Baker, Greaves and Wilson in the first two volumes of the *Observatory Publications*. Three further papers by Butler, Seddon and Thompson which appeared in 1959, 1960 and 1961 in the same *Publications*, brought this major Observatory programme to completion. Dr R. Wilson who had contributed significantly to the work, after a year's sabbatical leave at Victoria Observatory, Canada, was appointed at the end of 1958 to a post in the UK Atomic Energy Establishment at Harwell.

There can be no doubt that this Edinburgh spectrophotometric programme yielded data on spectral line intensities of the highest possible precision and that the measuring technique developed in Edinburgh – as Dr Wilson could show when comparing its results with those obtained at the Dominion Observatory, Victoria, by more standard methods – was capable of detecting spectral features such as the shallow interstellar absorption bands which would normally remain undiscovered.

However, the success of the Edinburgh work had been bought at the expense of quite an inordinate amount of labour and time jeopardising its effectiveness for the solution of topical scientific problems. While the basic principle of the Edinburgh technique of measurement had clearly proved itself, its efficient application provided an obvious example of a situation in which the introduction of fast automatic methods of measurement and reduction was demanded.

With both a limited staff and a modest budget at the time my ambition to branch out into the entirely new field of instrument technology had to await the suspension of other Observatory activities. The new venture became possible after the closing of the Observatory's solar work when Dr Ellison was appointed on 1st November 1958 to succeed me at Dunsink Observatory. Before leaving Edinburgh Ellison was able to complete his solar work with the installation at the Cape Observatory, South Africa, of a new Lyot H-Alpha Heliograph whose films of solar flares and active prominences, important for the programmes of the International Geophysical Year 1957–58, were for a while analysed in Edinburgh before all solar work and equipment was transferred to Dunsink.

Following this move it became possible to use the resulting vacancies and financial savings for the formal establishment within the Royal Observatory of an Instrumentation Group which was to become responsible for the technical support of most programmes in the Observatory and University Department. I was fortunate in being able to attract to Edinburgh to lead the work of the new group a scientist, Dr P. B. Fellgett, who had done much

The Royal Observatory East Tower and Library Wing.

pioneering work in Cambridge in the fields of optics, infrared astronomy and multiplex spectroscopy. One of the problems in which Fellgett was engaged at Cambridge at the time with the support of Professor R. O. Redman, was the feasibility of developing automatic methods for the analysis of photographs taken with Schmidt telescopes.

The remarkable properties of the type of telescope which Bernhard Schmidt had invented in Hamburg in 1930, had made it possible to photograph large areas of sky with high definition of stellar images. Telescopes of the Schmidt type became very popular, but their scientific value remained unexploited for a long time. Though highly praised for the remarkable pictures they provided particularly of nebulosities in interstellar space, Schmidt photographs were not sufficiently valued for the high content of information which they potentially contained.

When I was able to establish the new Instrumentation Group in Edinburgh under Dr Fellgett's direction it was agreed that the challenge of the effective use of Schmidt photographs ought to be taken up and that work on the design of a suitable Schmidt measuring engine should be amongst the group's top priorities.

The group consisting at first of only a small team of four, began its activities with the improvement of the efficiency of an iris plate photometer which was acquired by the Observatory at the time of Fellgett's appointment in September 1959. The photometer of the type introduced by Professor W. Becker of Basle recorded by means of a variable iris diaphragm a parameter embodying sizes and photographic densities of stellar images on photographic plates for the ultimate purpose of determining stellar brightnesses. The instrumentation group fitted that instrument with digitisers whose outputs representing both iris readings and stellar coordinates were automatically punched on paper tape in a form which could be used directly as data input to the EDSAC 2 computer of the Cambridge University Mathematical Laboratory, one of the few computers in Britain at the time. The acquisition of an in-house computer had to be shelved in those early days when computers of reasonable power were bulky and too expensive for the Observatory budget. In fact, the Observatory had to wait until 1966 before its first computer, an Elliott 4130, could be acquired.

The digitised iris photometer made it possible to put to proper use at last the observational material which had been collected in preceding years with the Edinburgh Schmidt telescope and whose analysis had been seriously hampered by the lack of suitable measuring equipment.

A substantial programme of research into galactic structure, based entirely on Edinburgh observations, was commenced by a group led by Dr V.C. Reddish who after three years at the Nuffield Radio Astronomy Laboratories, Jodrell Bank, had returned to Edinburgh as a member of the Royal Observatory staff. Another programme which made use of the digitised photometer in the early years, was carried out by Dr M. J. Smyth and Research Students in the University Department of Astronomy with the help of photographic material from the ADH (Armagh-Dunsink-Harvard) Schmidt telescope at Blomfontein, South Africa.

The final analysis of the measurements of the images of stars on Schmidt photographs depends crucially on the availability of independent photoelectric measures of the brightness of a range of standard stars in the same field with which the photographic data can be calibrated. This need arises from the well-known fact that unlike photoelectric detectors photographic emulsions do not respond in a linear way to the intensity of the radiation they receive.

In 1959 I was able to deal with this desideratum for our Schmidt work, the independent photoelectric measurement of a range of standard stars, by placing an order with Messrs Grubb Parsons in Newcastle for the supply of a telescope for photoelectric observations. This instrument which came into service in 1963, was the first new telescope on Blackford Hill since the installation of the Schmidt camera by Professor Greaves in 1951.

Experience in photoelectric photometry at the Dunsink Observatory had suggested that in conditions of variable sky transparency, high precision can still be achieved in photoelectric observations if the brightness of one star is measured in a single operation in terms of the brightness of a second star. On the basis of that experience the new telescope was designed as a twin instrument in which two identical Ritchey-Chretien telescopes of aperture 16 inches and effective focal length 20 feet are on the same mounting. Being moveable relative to each other over a certain range in both Right Ascension and Declination they can be pointed simultaneously at two different stars within five degrees of each other. The 16-inch apertures of the two telescopes were chosen to match that of the Edinburgh Schmidt telescope which the new instrument was intended to serve. Though mainly used in conjunction with the Schmidt telescope, the photoelectric telescope could also be used, of course, and was indeed used for

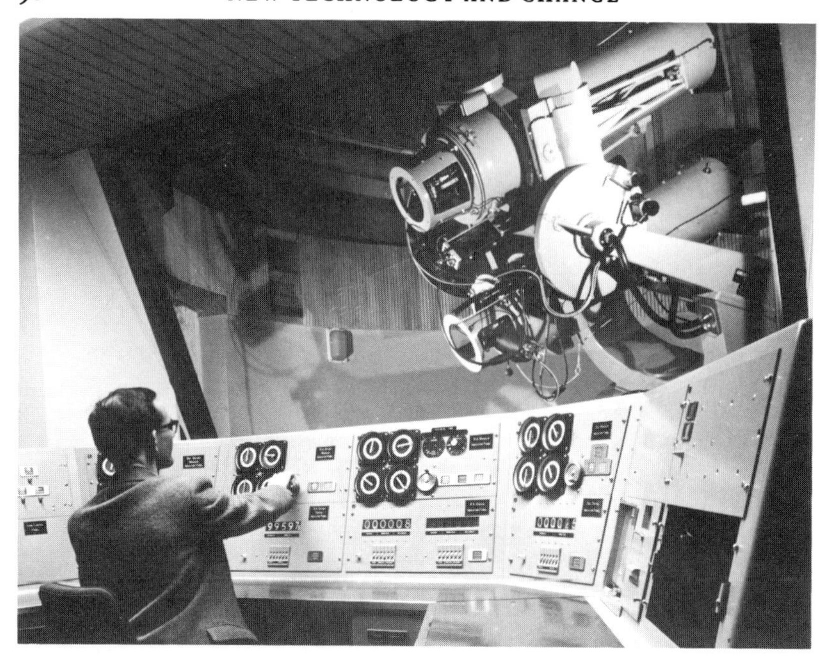

Top: The twin 16-inch telescope, installed in 1963, seen from its control room. *Bottom:* A Spectracon image tube attached to the spectrograph of the 36-inch telescope in the 1960s.

other purposes such as the monitoring of the light of variable stars and the testing of various new detectors.

The acquisition of this novel twin 16-inch telescope involved the instrumentation group in much further activity. While the telescope itself was built in Newcastle, the group undertook the design and construction of digitised photometers and programmable control circuits. Their work converted the telescope in the course of time into an instrument with which the brightness and colour of any star could be measured with precision and recorded on tape entirely automatically. Special efforts were made at the same time to create as perfect environmental conditions as possible for the telescope. This was achieved through the provision of natural and forced ventilation in the dome and through the separation of the observer from the telescope. The arrangements for remote control, now standard and in modern telescopes appreciably more sophisticated, were at the time the first of their kind anywhere and attracted considerable attention from astronomical instrument designers the world over. (This telescope has in the meantime been transferred to the University Observatory at St Andrews.) Some years later the instrumentation group provided similar control arrangements for the Schmidt telescope on Blackford Hill.

Amongst other instrumental innovations in those early years was the photoelectric exposure control for the spectroscopic work with the 36-inch telescope where the quality of spectrograms had been frequently affected by changes in sky transparency in the course of long exposures. Another was the provision of autoguider systems for both the 36-inch and twin 16-inch telescopes. These are devices which through photoelectric monitoring of a star's position in the field of view of a telescope automatically keep the instrument on target. Though once again autoguiders are standard equipment today, they were novel at that time when even in advanced institutions the old method of hand-guiding by the observers was generally used.

The early 1960s also saw experiments in the use of electronographic image tubes, specifically the Spectracon, developed by Professor McGee of Imperial College London. The attraction of the image tube for astronomers was its high sensitivity to faint light sources and its linear response to light intensity over an extended area. Tests of the device which started as part of a PhD project were carried out by Dr P. W. J. L. Brand who succeeded in securing a series of stellar spectrograms of high quality with the 36-inch telescope which later formed the basis of an important programme of research into the interstellar absorption bands in the spectra of distant stars.

By 1963 the planning by Dr Fellgett and his group of their most ambitious project, the fast automatic Schmidt measuring machine, was sufficiently far advanced that in August a contract could be placed by the Observatory with Messrs Ferranti Ltd, Dalkeith, for the actual construction of the machine. The development of this contract, the performance of the machine when tests were completed in 1969 and its further evolution will be described later.

In the early 1960s the 16/24-inch Schmidt camera had become the most widely used telescope on Blackford Hill which supplied data for a whole range of astronomical programmes. In addition to direct photographs of star fields, the telescope was also used to secure small-scale spectra of stars

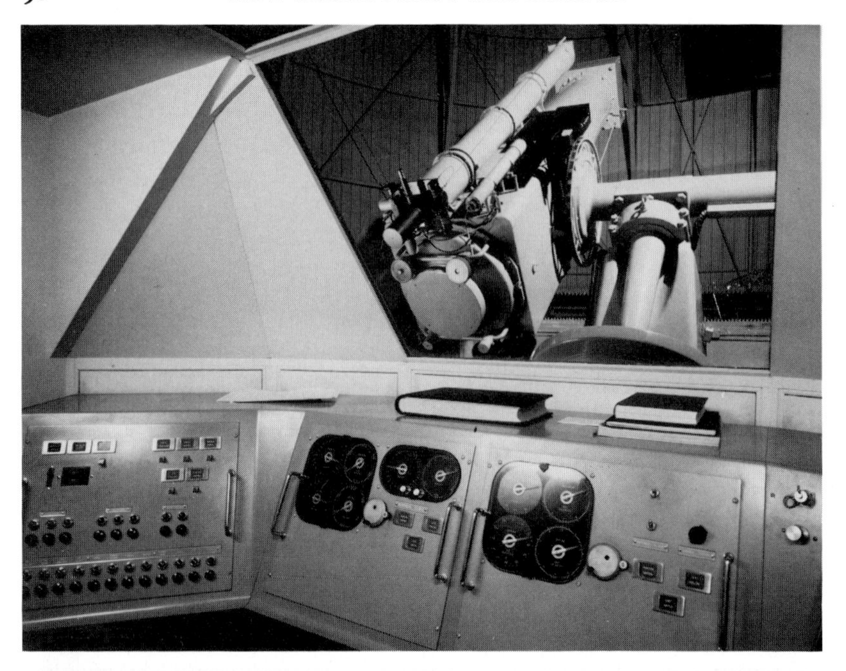

Top: The 16/24-inch Schmidt telescope and its controls.
Bottom: An objective prism and grating in front of the Schmidt telescope.

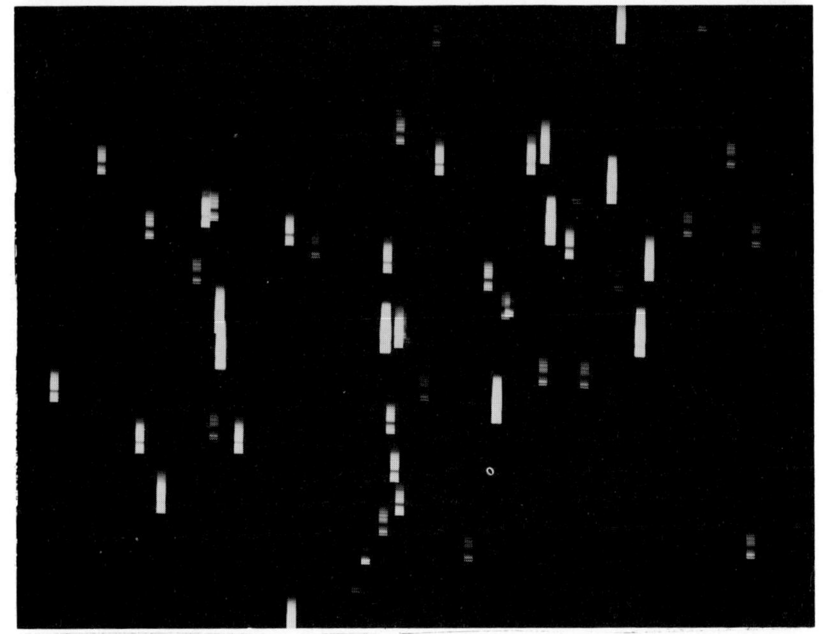

Top: The star cluster of the Pleiades. An early photograph taken with the Schmidt telescope on Blackford Hill. *Bottom:* Spectra of the Pleiades photographed with the Schmidt telescope through an objective prism.

by means of two objective prisms which had been constructed by Messrs Grubb Parsons and which gave spectral dispersions of 400 or 1000 Ångstroms per millimetre in the blue. The objective prism spectra were used by Dr K. Nandy for a substantial programme of research into the problem of interstellar obscuration.

At the same time we foresaw that Schmidt photography could not be carried out indefinitely from the Blackford Hill site where the effects of excessive brightness of sky and diminished atmospheric transparency were getting progressively worse as the City of Edinburgh expanded. A Schmidt telescope, being in fact a camera with small focal ratio, is particularly vulnerable to the effect of background sky illumination.

I looked at this situation as part of the general problem facing the pursuit of all optical astronomy in the climate of the British Isles. As far as Edinburgh was concerned, it seemed to call for the creation of an observing station in a more suitable climate, yet not too far away from Britain, where we could transfer our Schmidt work and in the longer term possibly other telescopes. From my own earlier experiences of observing at the Vatican Observatory at Castel Gandolfo and in view of happy relations with many Italian astronomers I considered Italy to be a particularly suitable place to start off a foreign project. Following discussions with Professors M. Cimino and L. Gratton of Rome University we received in 1963 an offer from Professor Cimino which I was delighted to accept of a site for an Edinburgh telescope within the extensive grounds of the astronomical observatory at Monte Porzio Catone near Frascati. Plans for a second, modern version of the 16/24-inch Schmidt telescope were then put in train and the new telescope was completed by Messrs Grubb Parsons and installed at Monte Porzio in 1967.

In the meantime an entirely new field of work was opened up at the Observatory as the result of the coming of the 'space age' and the creation of artificial Earth satellites. When the Russian Sputnik went up on 4th October 1957, Drs Butler and Reddish observed it from the roof of the Observatory on its first round, making Edinburgh the first optical observatory to record that historic event. It soon became clear that regular observations of the paths of artificial satellites were of considerable interest both for researches into the structure of the upper atmosphere and for geodetic triangulations. By 1960 the tracking of the various American and Russian satellites became one of the regular programmes of the Royal Observatory. It was carried out by a team of observers under the direction of Mr B. McInnes who joined the Observatory staff from the Physics Department of the University of Edinburgh. The observations were first made with cameras and a kinetheodolite loaned by the Royal Aircraft Establishment from the roof of the observatory on Blackford Hill. So promising and useful did this work prove to be that an outstation, exclusively for the tracking of satellites, was built in 1961 on a piece of land acquired by the Observatory at Earlyburn in Peeblesshire, 18 miles south of Edinburgh. The location, remote from interfering lights and with an unobstructed view of the horizon in all directions allowed the tracks of very faint satellites to be recorded photographically. This outstation, complete with laboratory and darkroom and living quarters for observers, was maintained for more than ten years during which it produced a com-

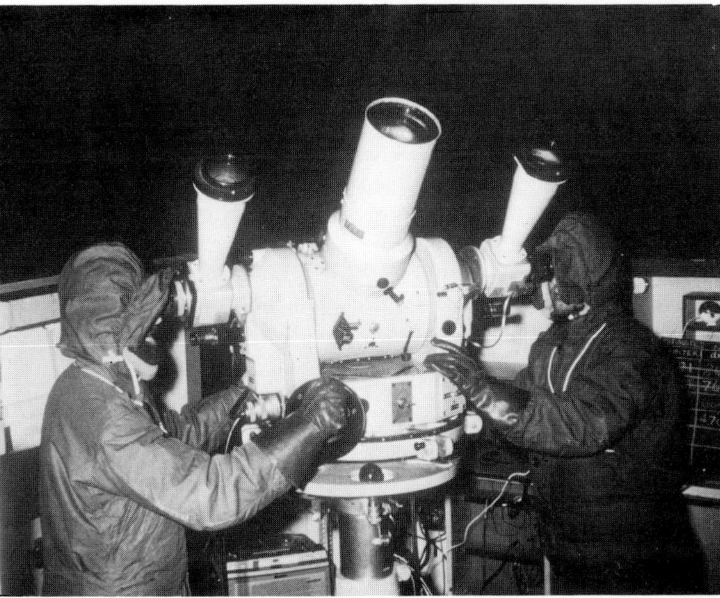

Top: Earlyburn Outstation in Peeblesshire, used for optical tracking of artificial Earth satellites from 1961 to 1975 and for testing of astronomical site-testing equipment. *Bottom:* Observers using the kinetheodolite for satellite tracking at Earlyburn Outstation.

plete record, weather permitting, of all observable satellites the number of whose transits increased from a few hundred to many thousands annually. All these Edinburgh observations were communicated to the World Data Centre at the Radio Research Station, Slough. Photographic recordings with the kinetheodolite were supplemented systematically by visual observations with binoculars and stop-watch.

Simultaneously with its satellite tracking programme the Royal Observatory took an early interest in the exciting new possibilities of making astronomical observations from outer space. This was at a time when British astronomers in general were reluctant to commit their limited resources to the pursuit of the new expensive field. In Edinburgh it was decided that some effort should be directed into work on the design and construction of astronomical instruments which could be flown first in rockets and later, possibly, in a satellite. A small Space Research Team was accordingly set up in 1961 under Dr H. E. Butler which consisted at the beginning of two Research Fellows, Messrs J. W. Campbell and G. C. Sudbury, who were later joined by a technical staff. The team became concerned with the design of instruments for direct and spectroscopic observations in the far ultraviolet part of the spectrum which were to be flown in British Skylark rockets from Woomera, Australia, and when in 1962 the United Kingdom had joined the European Space Research Organization ESRO, also from the rocket range at Salto di Quirra in Sardinia. The first scientific experiments, for which instrumentation was constructed at the Observatory, aimed at the observation of the brightness of the sky and of the energy fluxes and spectra of stars in the far ultraviolet.

While these experiments were in progress the Observatory was also able to embark on a collaborative programme with the Institut d'Astrophysique of Liège, Belgium, which was then under the direction of Professor P. Swings. This collaboration was concerned with the design and construction of a telescope, to be placed in a satellite which during a six months' period would scan the sky building up a picture of the radiation of stars, galaxies and general background as seen in four different wavelength ranges in the ultraviolet. In this cooperative effort the Royal Observatory became responsible for, amongst other matters, the design of the optics and structure of the proposed 28-cm telescope, for the calibration of the spectrophotometer and for the general assessment of the astronomical potential of the experiment.

The successful work of the Instrumentation Group and the existence of numerous new facilities in the Observatory was the cause of a further Edinburgh venture, the creation in the spring of 1962 of a Seismological Research Group. The Observatory had been responsible, of course, from its foundation for some geophysical work, and, as has been mentioned in an earlier chapter, a Milne-Shaw seismograph had been installed by Professor Copeland as early as 1900. This instrument had been in continuous use ever since, but sixty years later the question arose whether seismological activity was worth retaining in that form. An approach to the Royal Society of London resulted in their strong recommendation that the Observatory, far from abandoning its seismological work, should in fact use its expertise in instrumentation to modernise its activity in this field.

A Skylark rocket. The first British astronomical observations from space were made with equipment flown in Skylark rockets in 1961.

The chief target of the new group, which was led by Dr P. L. Willmore, formerly of Cambridge, and which eventually was to grow to seven members, was the design and construction of equipment for recording and playing back seismic data on multichannel magnetic tape. To assist the development work of the Group which remained closely associated with the Observatory and the Astronomy Department for several years, a large subterranean vault was constructed under the western part of the main Observatory building which incorporated various special facilities for new seismological instrumentation. A set of directive aerials was also installed for the reception of seismic signals from outlying field stations.

The success of the Seismology Group's activities led to the establishment in 1964 of an internationally financed International Seismological Centre set up under the auspices of Edinburgh University. To cope with this expansion the University put at the disposal of the Seismology Group a property close to the Observatory at South Oswald Road which became the headquarters of the new Centre.

Following the installation of the new seisomological equipment the Observatory's old Milne-Shaw seismograph was donated to the Royal Scottish Museum where it was set up for public display. The Museum was presented at the same time with the last of the Observatory's pendulum clocks, the Molyneux mean-time clock which had been used continuously since Calton Hill days for the daily time service of the City of Edinburgh and in earlier years also of that of the city of Dundee. The two other pendulum clocks of

Lady Tweedsmuir at the inauguration of the twin 16-inch telescope in 1963.

the Observatory, Leroy No. 1230 and Riefler No. 258, had been handed over to the Museum in 1959 when there was a general change-over from pendulum to quartz clocks for the Observatory's timekeeping.

The numerous new activities of the Observatory resulted in the need for more space and already in 1958, a year after my arrival, HM Ministry of Works as it then was drew up plans for workshops, laboratories and a dome for the new proposed 16-inch telescope. The extensions were to be to the west of the main Observatory building and in the place of the house for the old transit circle which was to be dismantled.

Building began in 1959 on a new workshop to replace the Observatory's original workshop in the half-basement beneath the main library, which in turn released much needed accommodation for a library annexe. This was followed by the building of the dome for the new 16-inch twin telescope and a year later by a two-storey laboratory and office block erected between the workshop and the main building. This block contained a lecture room, an optical laboratory later to house the Observatory computer, an electronics workshop and a number of smaller rooms. The design of these buildings was the responsibility of Mr J. McMinn, Architect of HM Ministry of Works, who kept very much in mind their effect on the skyline of Blackford Hill as viewed from the City of Edinburgh and who also took great care to blend as happily as possible the new with the old building. The extension had a flat roof equipped with stone benches for portable instruments for undergraduate teaching sessions; it was also used and is still used to carry a set of aerials connected to the seismological equipment in the vault underneath.

Improvements in the old building included the conversion of the clock cellar in the basement into a vacuum laboratory to contain an aluminising plant for the mirrors of the Observatory's telescopes and equipment for the testing of rocket instruments.

By the time the new workshop was ready, Mr G. J. Matthews, the Resident Mechanic, could be joined by an increasing number of engineering staff.

In 1963 the modernisation of the Royal Observatory was sufficiently far advanced to allow the various new facilities and, in particular, the new 16-inch twin telescope to be opened officially. It also seemed an appropriate time to review the progress of research at the Observatory, and we took the opportunity of doing this in connection with a meeting of the Royal Astronomical Society held in Edinburgh in September of that year. Two international scientific colloquia were organised, one on 'Problems of the Interstellar Medium and Galactic Structure' and the other on 'New Methods in Seismology'.

The formal opening ceremony was performed by the late Lady Tweedsmuir, who was then Joint Parliamentary Under-Secretary of State for Scotland. The guests on that occasion included the late Sir Edward Appleton as Principal of Edinburgh University and Sir Richard v.d.R. Woolley as President of the Royal Astronomical Society.

In her Address Lady Tweedsmuir stressed 'the link between Government research and University interest which gives the Royal Observatory unique opportunities to advance knowledge of astronomy and at the same time to train a new generation of astronomers'. This theme, the close, happy and unique connection between the Observatory and the University of Edinburgh was further emphasised by Sir Edward Appleton when he said in his Address: 'I have never heard of any demarcation here at Blackford Hill, and I do not suppose anybody knows which is the University Department and which is the Royal Observatory. I suppose, all of it is both.'

A detailed account of the proceedings at the formal Opening and of the various astronomical scientific papers and discussions was published in the fourth volume of the *Observatory Publications*. The seismological discussions were reported in the fourth volume of the *Quarterly Journal of the Royal Astronomical Society*.

12

OUTSTATIONS AT HOME AND ABROAD

THE YEAR 1965 was a particularly memorable one in the annals of the Observatory. On 1st July the Observatory was honoured by a visit from Her Majesty The Queen and His Royal Highness, The Prince Philip, the first Royal visit since its foundation. Earlier in the year an important step with far-reaching consequences had been taken when as a result of the reorganisation of Civil Science in the United Kingdom the Royal Observatory became an establishment of the newly created Science Research Council (now the Science and Engineering Research Council).

The Royal visit had been carefully planned and everything was done to ensure that the scientific and technical work of the Observatory should be displayed in as interesting a manner as possible for this unique occasion. The Royal Party made a detailed tour of the Observatory, inspecting instruments, models and displays illustrating the various branches of the Observatory's work. They were conducted by Sir Harry Melville, the first Chairman of the Science Research Council, and myself, and met members of staff and a group of distinguished guests with particular links with the Observatory. Among them were Sir Michael (now Lord) Swann, then Principal of Edinburgh University, the Earl of Crawford and Balcarres, Mr George Sisson, who was then Managing Director of Grubb Parsons Ltd, the Newcastle telescope makers, and Dr E. A. Baker, then a retired member of staff and the oldest and longest serving of the Observatory's astronomers. At the end of her visit Her Majesty was presented with and graciously accepted for the Library of Windsor Castle a miniature model of the new Schmidt telescope which was under construction at the time for the Observatory's outstation in Italy.

For two days after the Royal Visit the Observatory was open to the general public when more than a thousand people took the opportunity of viewing the displays on the Observatory's work.

The change-over of the administration of the Observatory from the Scottish Office to the Science Research Council involved much detailed discussion since we in Edinburgh were determined to retain the unique character of the Observatory which links it to the University of Edinburgh. I had always looked at this arrangement which identifies the Astronomer Royal for Scotland with the Regius Professor of Astronomy and which dates back to the Observatory's foundation, as one of the particular strengths of the institution. I was fortunate to be always supported in this view by the Principal of the University, Sir Edward Appleton, who took a very sympathetic interest in the Observatory's affairs and who, while pointing out the possible advantages of central funding for large future astronomical requirements, strongly encouraged my view on the advantages of the particular position in Edinburgh.

The actual transition was helped by the fact that Sir Harry Melville had known the Observatory well in his student days at Edinburgh University and was very supportive of our efforts. Though I was well aware of the benefits which would result ultimately for the Observatory from the new arrangements, I could not help regretting the ending of a long and happy association with the Scottish Office. At St Andrew's House in Edinburgh I had found great understanding for my often unusual ideas and I had been allowed to present my annual requests for financial support of the Observatory's work direct to HM Treasury in London with the result that problems could be discussed and settled with the minimum of delay. A major part of the modernisation and expansion of the Observatory had in fact been achieved under the aegis of the Scottish Office with the scientific and technical staff having increased from seven to well over forty at the time of

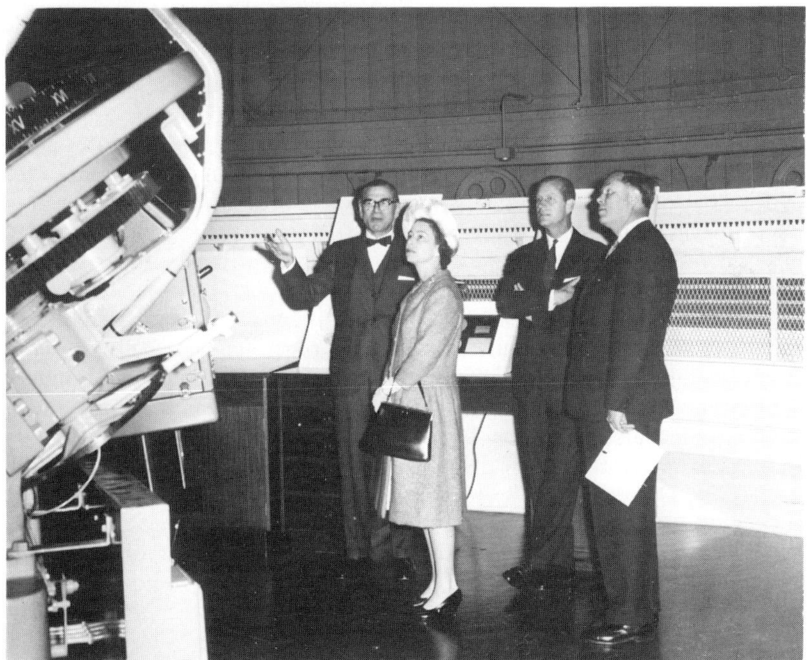

Top: Her Majesty the Queen visits the Observatory on 1 July 1965.
Bottom: H.M. The Queen, H.R.H.Prince Philip, and Sir Harry Melville, first Chairman of the Science Research Council, are shown the 36-inch telescope.

An iris plate photometer operated on-line to the Observatory's computer which came into use in 1966.

the transfer of administration to the Science Research Council.

In coming under the aegis of the Science Research Council the staff of the Observatory who had previously belonged to the Scientific Civil Service had to take out new contracts of employment with the Council. The executive and clerical personnel of the Observatory had traditionally served on secondment from the Scottish Office; some of these members of staff, realising that under the new contracts they would be liable to be transferred to London or other SRC establishments south of the Border, preferred to return permanently to St Andrew's House. In this way the Observatory had to say goodbye to some very efficient office staff who like their predecessors had served the Observatory well.

I paid my last official visit to St Andrew's House in March 1965. From then onwards the new venue of numerous meetings was State House in High Holborn, the then Headquarters of the Research Council in London. The changed circumstances inevitably involved additional administrative work and the Observatory was very fortunate in being able to welcome in December 1968 as Deputy Director Dr R.H. Stoy, formerly HM Astronomer and Director of the Cape Observatory. He also became an Honorary Professor in the Astronomy Department of the University and continued in these two capacities until his retirement in 1975. Among the various Boards and Committees which were set up by the Council was a Committee for the Royal Observatory Edinburgh which met once a year at the Observatory to advise on the Observatory's activities and future plans. This Committee met for the first time in January 1966 with Professor R.O. Redman as its Chairman. Being mainly composed of University scientists this Committee took a very positive interest in the Observatory's work which continued when in 1970 Professor Redman was succeeded as Chairman by Professor D.W.N. Stibbs of St Andrews.

The year 1965 also saw the appointment as Professor of Cybernetics at Reading University of Dr P.B.Fellgett who in his six years at Edinburgh had been the inspiration of many of the innovations which had been pursued at the Observatory with considerable success. Work on the construction of the fast automatic measuring machine for Schmidt photographs which Fellgett had called 'GALAXY' (from *G*eneral *A*utomatic *L*uminosity *A*nd *XY* measuring machine) was well on its way at the time, but there were difficulties and delays before the machine was finally delivered in 1967 by Messrs Faul-Coradi Scotland Ltd who had taken over from Messrs Ferranti Ltd, Dalkeith. The final installation of GALAXY at the Observatory was the joint work of Fellgett's successor as Head of the Instrument Group, Mr G.J.Carpenter, a senior Engineer, of Dr V.C.Reddish and of Mr G.S. Walker of Faul-Coradi Ltd. In the same year 1967 the Observatory also received its first computer, an Elliott 4130.

While awaiting the installation of GALAXY, members of the group had experimented with the design of smaller semi-automatic instruments. When in 1965 a second iris plate photometer identical with the first one which was in constant use, was acquired by the University Department, the instrument was adapted for on-line operation with the Observatory's new computer. It became then possible to program work of the photometer by supplying the computer with the coordinates of stars on a 'master' photographic plate allowing subsequently each plate of the same field to be measured entirely automatically and iris readings and coordinates of each star to be recorded on punched paper tape. The associated software allowed the data from all plates of the same field to be calibrated and combined into a list of brightnesses and colours for every star in the field. This semi-automatic plate photometer represented a major innovation at the time though it was relatively slow, of course, and could only serve as an interim solution of the problem of the analysis of Schmidt photographs. Before the arrival of GALAXY it played a very useful part, however, in the measurement of large numbers of Schmidt photographs.

Other work of the Instrumentation Group included the digitisation of a Hilger recording microphotometer and the preparation of a programme of automated stellar spectrophotometry which was to preserve the advantages in precision of the earlier work of Professor Greaves, Dr Baker and associates without suffering its disadvantages in speed.

The GALAXY machine arrived at the Observatory in 1967 and was installed in specially controlled surroundings in a room within the new wing which had just been built for the University Department that same year. It took nearly two years before a series of tests of the exact performance of this highly complex machine was completed to the satisfaction of the Observatory. The original purpose of GALAXY had been to find the images of stars on a photographic plate and, having found them, to measure their positions and to record, as measures of their brightnesses, their sizes and photographic densities. These operations were to be carried out with high speed and precision completely automatically. According to the original specification positional measurements were to be accurate to a micron and brightnesses were to be recorded with a precision of two per cent. The actual performance of this remarkable machine turned out to be even better. In its

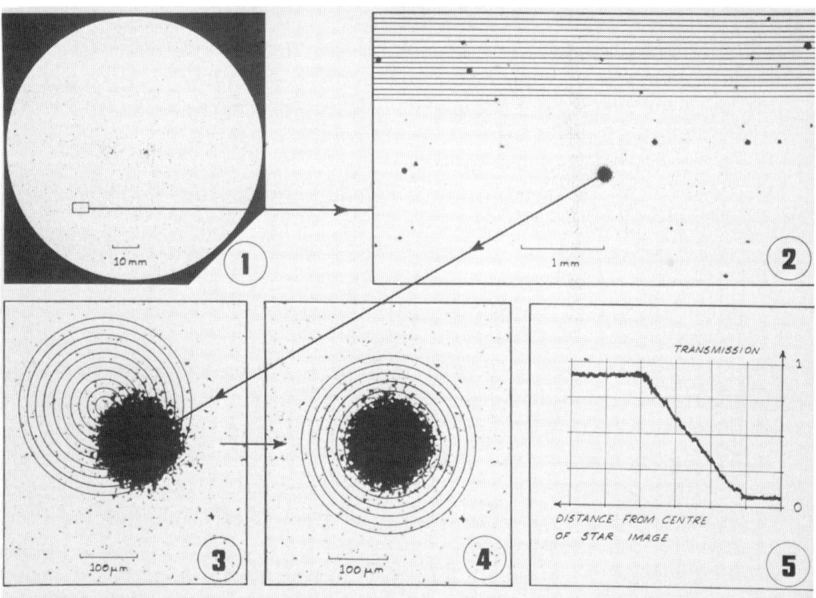

Top: The automatic plate measuring machine GALAXY, 1967. The photographic plate is mounted horizontally for scanning behind the door, shown open in the photograph. *Bottom:* The scanning process employed in the 'search' and 'measuring' modes of GALAXY.

'search mode' GALAXY could record the approximate positions of images of stars at the rate of 10,000 stars per hour and in its 'measuring mode' it could print the precise positions of the same stars at the rate of 900 stars per hour with an accuracy of better than a micron.

The completion of GALAXY marked a milestone in the Observatory's work on advanced measuring facilities. The speed and precision of the machine were such that for the first time programmes could be contemplated which demanded the measurement of tens of thousands of stellar images or the detection of particular types of stars amongst a host of others. Dr Reddish did not exaggerate when he declared that the creation of GALAXY heralded a new era in astronomy.

The new machine was used for the analysis of large numbers of Edinburgh Schmidt photographs, but it was also placed at the disposal of astronomers and other scientists from Britain and abroad. A second GALAXY, a copy of the first machine, was in due course installed at the Royal Greenwich Observatory. While the first GALAXY was chiefly used for photometric programmes in astrophysics, the second has been concerned mainly with astrometric measurements in positional astronomy.

The success of the GALAXY machine as well as of the smaller automated instruments constructed at the Observatory prompted us to invite the International Astronomical Union to hold a formal Colloquium in Edinburgh on the topic of 'Automation in Optical Astrophysics'. The three-day meeting in August 1970 was opened by Professor J. Rösch, the French astronomer and Chairman of Commission 9 on Astronomical Instruments of the International Astronomical Union, and was attended by eighty astronomers and instrument physicists from all over the world. A total of forty papers were read on topics of telescope and auxiliary instrument control, on automated measuring equipment and on data handling and reduction. Two definitive papers on the GALAXY machine, one on its design and development by Mr G. S. Walker of Faul-Coradi Ltd, the firm which completed the machine, and the other by Dr N. M. Pratt of the Royal Observatory Edinburgh on the machine's performance. Another definitive paper was presented by Mr G. Adam of the Observatory who described the autoguider which he had designed and which had been incorporated in Edinburgh telescopes. The Edinburgh autoguider became a standard accessory on other UK telescopes and was later manufactured for wider distribution under license by Messrs Grubb Parsons in Newcastle upon Tyne.

The proceedings of the Automation Colloquium (edited by H. Seddon and M. J. Smyth) were published in the *Publications of the Observatory*. They are by now a historic publication representing as they do the exact state of the art at the time. The success of the Colloquium owed a great deal to its Edinburgh venue where much had already been accomplished and where participants could see in actual operation projects which elsewhere were only at the planning stage.

While GALAXY was being installed in Edinburgh the Observatory's new outstation in Italy came into full operation. Its main instrument, an up-to-date version of the Observatory's 16/24-in telescope, was mounted in a building which had been leased from the Rome Observatory. Its site was close to the village of Monte Porzio Catone and the ancient Tusculum, the

Top: View of the Royal Observatory in 1967 showing the original buildings and the modern extensions with the domes of the 16-inch twin telescope (*left*) and the University's 20-inch telescope (*background*). *Bottom:* The Royal Observatory Edinburgh Outstation at Monte Porzio Catone, Italy, showing the telescope dome in the foreground and the building containing the Michelson interferometer to the right.

The 16/24-inch Schmidt telescope at Monte Porzio.

birthplace of Cato the Elder and a favourite resort of Cicero. The Vatican Observatory at Castel Gandolfo was only a few miles away and the Italian Space Laboratories in Frascati were also close by.

The Rome Observatory complex at Monte Porzio including the building for the Edinburgh telescope had had a somewhat bizarre history. The imposing structures had been intended by Mussolini's Government to house a majestic Italian National Observatory. According to Father J. Stein, the Director of the Vatican Observatory in the 1930s, the idea had originated in 1938 in the course of a visit paid by Hitler to Mussolini. There was an exchange of gifts in which Hitler promised Mussolini to provide all the telescopes for the creation of an observatory which would rival in every way the then new Vatican Observatory at Castel Gandolfo. However, when the Second World War had broken out the erection of the buildings was soon suspended and ultimately abandoned and the telescopes and their domes were removed by the German army at their retreat from Italy in 1944. It was not until the early 1960s that Professor Cimino could consider the reconstruction of some of the various buildings of the originally proposed new observatory. The completion of the building which was to house the new Edinburgh Schmidt telescope became a major part of this reconstruction. When completed, the Royal Observatory Edinburgh building provided rooms for offices, workshops and laboratories, a small library and sleeping accommodation for two observers. The dome, 20 feet in diameter, for the Schmidt telescope had been secured from an Italian firm, Messrs Piermattei of Rome.

This attractive outstation was opened formally on 26th October 1967 by Sir Evelyn Shuckburgh who was then British Ambassador to Italy, in the presence of the then Italian Minister for Education, Professor L. Gui, and a large contingent of British and Italian scientists. It was an occasion for an exceptionally happy display of Anglo-Italian co-operation which in fact resulted in much joint astronomical research between various members of the Royal Observatory staff and Italian astronomers working at Monte Porzio, Monte Mario and the Rome Astrophysical station at Campo Imperatore on the Gran Sasso of Italy.

The presence of astronomers from Scotland was not lost on some of the local dignitaries who were aware of the romantic historical connection between Scotland and Italy through Prince Henry, Duke of York, brother of Bonnie Prince Charlie and last of the House of Stuart, who had been a very much-loved Bishop of Frascati.

The new outstation was usually manned by two observers from Edinburgh who were assisted by an Italian electronic engineer, Signor F. Rossi (who is now on the staff of the Vatican Observatory). The station became responsible for large numbers of Schmidt photographs, taken directly or with one or the other of objective prisms, and it produced much observational material for the work of Research Students in the University Department of Astronomy.

In 1969 the Schmidt Telescope at the Station was joined by a 2-metre Michelson Interferometer which had been previously developed at the National Physical Laboratory in Teddington. The astronomical purpose of this interferometer was to measure the smallest possible angular separation of objects such as close double stars and even to determine the finite diameters of exceptionally large single stars. The instrument was under the direction of Dr R. Q. Twiss who had been appointed an Honorary Research Fellow of the Royal Observatory and who had developed the instrument to a stage where the effects of atmospheric interference on the observations could be measured numerically. In the pursuit of the observation of close double stars a second improved version of the interferometer was completed in the Observatory workshops in 1972. The project in which Dr Twiss was assisted by Dr W. J. Tango as a Research Fellow and Mr K. Russell of the Royal Observatory had not been completed when the Monte Porzio station was closed in 1976.

While work at the Italian outstation was progressing Edinburgh efforts in the field of space research were coming to fruition. Following earlier disappointments the first Edinburgh rocket firings took place successfully in 1965 and were followed by a series of others over the next six years which altogether yielded photometric and spectrophotometric data on the radiation of stars and sky background in the ultraviolet down to a wavelength of 1500 Å.

In 1966/67 the design study was completed of the instrumentation for the joint Edinburgh/Liège experiment which was to be flown in the ESRO satellite TD1. Because of the complexity of the instrumentation and the high expenditure involved the management of this satellite project was reorganised in 1968 when a special Committee with Professor R. Wilson as Chairman, Dr Butler as Project Astronomer and two members from outside the Observatory in addition to the Edinburgh staff, took over the supervision of the work. By 1971 tests gave clear evidence that the Edinburgh equipment

Photograph taken at the opening of the Monte Porzio Station showing, in the centre, Sir Evelyn Shuckburgh with (*right to left*) Professor M. Cimino of Rome University, Signor L. Gui, Italian Minister for Education, Dr W. L. Francis of the Science Research Council and the author.

worked entirely satisfactorily and also that the necessary data processing was well in hand. A year later, in March 1972, the TD1 satellite was successfully launched into orbit. It remained active for two years and the observational data received from its survey of the sky led to a substantial number of measurements of ultraviolet fluxes for several thousand stars and to the joint publication of a major catalogue by the Edinburgh, Liège and London teams involved in the work.

The Observatory's activities in the field of satellite tracking received a considerable boost when in 1966 a large 24/36-inch, f/1, Hewitt Schmidt Camera was moved to the Earlyburn outstation from the Radar Research Establishment in Malvern. This camera proved capable of timing the transits of satellites with an accuracy of a thousandth of a second and of providing angular positions with a precision of one second of arc. For a while this camera was joined by equipment of the United States Coast and Geodetic Survey. Work on satellite tracking had to be curtailed in the early 1970s when members of the tracking group, on account of their observational experience, were required to serve in a new site testing project for a national observatory overseas. Some work continued, however, for a number of years, thanks mainly to the enthusiasm of Mr R. Eberst who was later to receive the Merlin Medal of the British Astronomical Association for his work.

Following the general re-organisation of civil science under various new Research Councils the seismological activities of the Royal Observatory were transferred to the Institute of Geological Sciences of the Natural

The 36-inch Hewitt Schmidt camera for rapid photography of artificial Earth satellites installed at Earlyburn Outstation in 1966.

Environment Research Council. However, the large seismic vault constructed under the 1963 building on Blackford Hill continued to be used up to the present time by the 'Global Seismology Unit' of the Institute.

In 1967 the Observatory ventured also into the relatively new field of infrared astronomy. Scientists in Imperial College London had already set up a 60-inch flux collector for infrared work at Izaña on the island of Tenerife. The Edinburgh team consisting of Dr M. J. Smyth of the University Department and Messrs G. Cork and J. Harris of the Observatory began, in collaboration with the Instrumentation Group, the development of various detector systems which were later used in Tenerife, at the Isaac Newton Telescope at the Royal Greenwich Observatory, and at the 74-inch Radcliffe Telescope in Pretoria, South Africa. The Pretoria observations, using a rapid scanning Fourier spectrometer, were in fact the first infrared spectroscopic observations made in the southern hemisphere. The instrumentation group, under Mr Carpenter, was also involved in improvements of the control system of the 60-inch flux collector on Tenerife. In this connection mention should be made of the electronic engineers who took part in this and many other projects, Messrs D. H. Beattie, R. J. Beetles, R. W. Parker and T. E. Purkins.

The celebrations surrounding the 150th Anniversary of the Royal Observatory in 1972 included a Colloquium on Infrared Astronomy in the United Kingdom which was attended by representatives of all British groups active at the time. The proceedings of this meeting (edited by

M. J. Smyth and H. Seddon) were published in the 9th volume of the *Observatory Publications*, the last in the series.

13

THE ERA OF NATIONAL FACILITIES

THE RE-ORGANISATION of British Science under the Science Research Council brought with it the promise of new opportunities for astronomers by the creation of major national facilities which would be shared by research workers in the universities and the Royal Observatories. It seemed now at last possible that British astronomers would in due course have at their disposal powerful telescopes and other equipment which would allow them to observe over the whole spectrum faint and distant objects which so far had been largely the prerogative of their American colleagues. Such large instruments were bound to be costly, of course, and neither the Royal Observatory Edinburgh nor any other single establishment could any longer expect to be provided with the necessary expenditure of public money. The new instruments had to be created as national facilities.

The first major telescope to be embarked on under the new dispensation was a 150-inch reflector, the Anglo-Australian Telescope, whose construction had been proposed by the Government of Australia to the British Government in 1967 on the understanding that the two Governments would share the cost of both the construction and operation of the instrument which would be placed in Australia. The proposal which was strongly supported by Sir Richard Woolley who was then Astronomer Royal, was accepted by the Science Research Council for the British Government and it was decided that the new telescope should be sited on Siding Springs Mountain in New South Wales.

The experience gained in preceding years by the Edinburgh Instrumentation Group in work on telescope control proved important when it came to design work on the control of the new telescope. Both Mr G. J. Carpenter and Mr T. Wallace of the Royal Observatory staff became closely associated with this work and in 1971 Mr Wallace actually joined the Anglo-Australian Telescope Unit in Australia as a computer systems analyst. His sudden death in 1974 shortly before he was due to return to Edinburgh, was a grievous loss to the Royal Observatory as well as to the Australian team.

The proposal of the creation of an Anglo-Australian Telescope had been given priority over a proposal which I had made at about the same time for the construction of a large telescope in the northern hemisphere. This was to be a 150-inch telescope to be placed at a carefully selected site at a reasonable distance from Britain where the number of clear nights and, even more important, the definition of stellar images was as perfect as possible.

The first paragraph of the submission which I made to the Science

Research Council in January 1967, contained the following arguments:

> Lack of major optical facilities excludes British astronomers at present from important areas of research where emphasis moves towards investigations of fainter objects in the Galaxy or of more detailed spectral analysis, and towards studies of external galaxies and quasi-stellar objects. The proportion of exciting astronomical research grows in which British optical astronomers cannot participate without using the hospitality of foreign colleagues, and this situation is all the more deplorable as present advances in astrophysics stem in no small measure from the brilliant work being done by British radio astronomers and theoretical astrophysicists.

The Advisory Committee appointed by the Science Research Council for the Royal Observatory Edinburgh strongly endorsed that proposal in February 1967, but the scheme for the Anglo-Australian Telescope was given priority at the time. It may be worth noting that at their meeting in March 1969 the Royal Observatory Committee recorded their view that 'it might well be possible to build a Northern Hemisphere Telescope after the Anglo-Australian Telescope comes into service, starting during the period 1974/79 with completion in the early 1980s'.

In March 1969 I submitted to the Science Research Council a revised version of my original proposal of January 1967 in which I now suggested, with a preliminary costing exercise, that the proposed 150-inch telescope should be accompanied by some smaller instruments such as a 2-metre and 1-metre telescope which together would constitute a 'Northern Hemisphere Observatory'.

The idea of the creation of such a major British observatory in the northern hemisphere was taken up and strongly endorsed by the 'Northern Hemisphere Review Committee' which had been set up by the Research Council under the Chairmanship of Sir Fred Hoyle. Its purpose was to review the general state of British astronomy in the northern hemisphere and to suggest possible improvements. The suggestions by the Committee followed those of an earlier review by the Science Research Council of the state of British optical astronomy in the southern hemisphere, a review which had led to the establishment of a new South African Astronomical Observatory equipped with telescopes of the former Royal Observatory at the Cape and the Radcliffe Observatory in Pretoria.

The Northern Hemisphere Review Committee had its first of many meetings in January 1969. It soon came to the conclusion that the proposal for the creation of a Northern Hemisphere Observatory should be supported and that serious efforts should be made to identify the best site for such an observatory. The systematic testing of possibly suitable sites started in 1970 as a joint project of the two Royal Observatories, but in April 1971 full responsibility for the site-testing project was taken over by the Royal Observatory Edinburgh. Mr B. McInnes was put in charge of the whole operation, whose staff consisted of a few permanent members of the Observatory staff and some short-term temporary assistants who were recruited specifically for this work. The Observatory workshops constructed the necessary equipment consisting of identical Polaris-trail telescopes and sky brightness/transparency monitors for the use by the various ex-

A member of the Edinburgh site-testing team at work at Fuenta Nueva on La Palma, the actual location finally chosen for the new Northern Hemisphere Observatory.

peditions which were sent out from Edinburgh to a number of possible locations.

The choice of these locations followed examination of meteorological records and of results of site-testing operations carried out by French, German, Italian and Spanish teams who were looking for suitable sites for new observatories of their own. Two of the three regions finally chosen by the Edinburgh group were in southern Italy and in the Canary Islands where the very first site-testing operation had been carried out on the island of Tenerife by Professor Piazzi Smyth of Edinburgh in 1856. A third region originally contemplated by the Edinburgh team in south-east Spain proved inaccessible to them.

Tests of the chosen Italian site in the Province of Potenza – since revealed as liable to major earthquakes – which were carried out between February and November 1972, were not particularly encouraging. In contrast, observations made between December 1971 and December 1972 on the neighbouring Canary Islands of Tenerife and La Palma revealed very promising conditions there with a distinct advantage of La Palma over Tenerife as far as astronomical seeing and level of artificial lighting were concerned. The site-testing team also studied conditions on the island of Fogo in the Cape Verde Archipelago and on the highest point of the island of Madeira. However further observations on the La Palma site made in 1974 and 1975 confirmed the earlier assessment of the excellence of the site and the Planning Committee made the decision to choose Fuenta Nueva or its highest point, Roque de los Muchachos, on the island of La Palma as the site for the proposed new observatory.

While the Anglo-Australian Telescope was being constructed in Australia, the Science Research Council came to the decision that British optical

Top: The UK 48-inch Schmidt telescope at Siding Springs, New South Wales, Australia, completed in 1973. *Bottom:* Dome and buildings of the UK 48-inch telescope at Siding Springs.

astronomy would also greatly benefit from a large wholly British Schmidt Telescope which would operate from the same site at Siding Springs in New South Wales. It was agreed in 1970 that this new telescope should be an up-to-date version of the famous 48-inch Schmidt telescope on Mount Palomar in California which had come into operation in 1948 and which had produced the well-known survey of the northern skies. The responsibility for the planning and execution of this major project was entrusted to Dr V.C. Reddish as Project Manager who assembled a small unit which included Dr R.C. Cannon and Miss M.E. Sim with headquarters at the Royal Observatory Edinburgh.

The telescope built by Messrs Grubb Parsons in Newcastle was completed in record time. By 1973 it was already in full operation producing large numbers of deep-sky photographs of a quality never previously achieved. The telescope, known as the UK Schmidt Telescope (UKST) had as its primary aim a photographic survey of the southern skies and within two years of that date a large fraction of this task was already accomplished as well as the supply of hundreds of additional photographs made available for special research programmes.

In 1974/75 the Edinburgh site-testing work was extended to sites on the summit area of Mauna Kea in Hawaii in preparation for the proposed mounting there at an altitude of 14,000 feet of a second large British telescope which was to be devoted specially to observations in the infrared. Proposals for the construction of a major 'Infrared Telescope' had been put before the Science Research Council as early as 1968 and following the successful operation of the 1.5-metre prototype telescope mounted at Izaña on Tenerife by a team from the Imperial College of Science and Technology in London – using a control system designed and built at the Royal Observatory Edinburgh – a large infrared telescope with an aperture of 3.8 metres was proposed and finally approved by the Department of Education and Science in June 1974. The arrangements for this new project of a large Infrared Telescope were such that Professor J. Ring of Imperial College was appointed Project Scientist and Chairman of a Steering Committee composed of scientists involved in infrared work, and that Mr G. J. Carpenter, the Head of the Royal Observatory's Instrumentation Group, was appointed Project Manager. Another senior member of the Observatory staff, Dr T. J. Lee, was at the same time made responsible for the provision of common user instrumentation for the new telescope. The Edinburgh staff was therefore taking a major part in the creation of the new 'United Kingdom Infrared Telescope' (UKIRT). The Observatory was also made responsible for the operation of the telescope on the Mauna Kea site which like the UKST was to become a national facility available to all competent astronomers in the United Kingdom.

When as the result of the coming into being of the new 48-inch Schmidt telescope researches at the Royal Observatory could be extended from the field of galactic to that of extragalactic astronomy, it became desirable to widen the scope of the measuring engine GALAXY. The machine was to be able to cope with the measurement of non-circular images of galaxies and other objects as well as with that of circular images of stars. When financial restrictions prevented the construction of a second independent GALAXY-

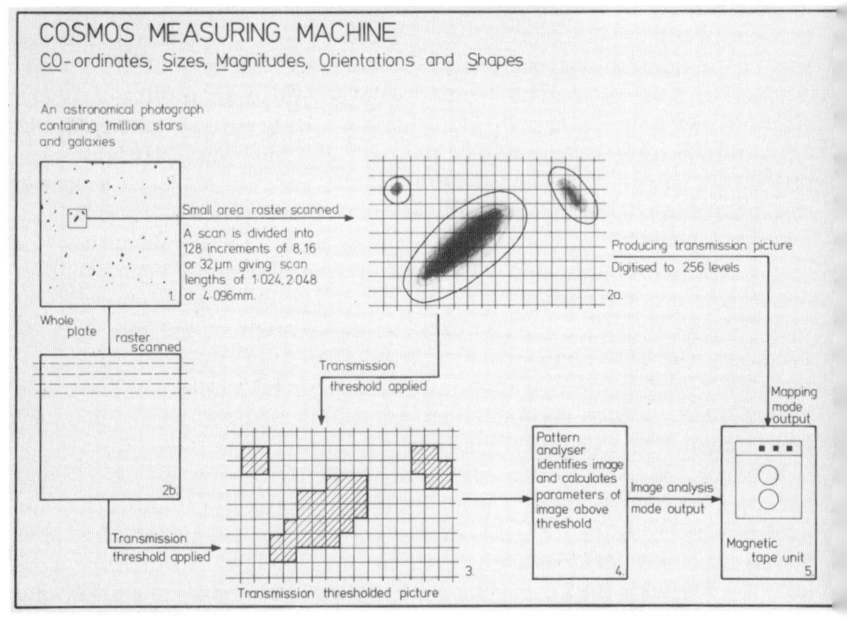

Top: The automatic plate-measuring machine COSMOS, 1975.
Bottom: The scanning process employed in COSMOS, which recognises extended as well as circular stellar images.

The 150-inch UK infrared telescope on Mauna Kea in Hawaii. The project was started in 1974 and completed in 1979 under the direction of Professor V.C. Reddish.

type machine for Edinburgh it was resolved – with some regret – that the original GALAXY machine should be redesigned so that it could deal with the new problem. The task of making the necessary alterations was entrusted to Dr V.C. Reddish who with the co-operation of Messrs Faul-Coradi Ltd and a team from Heriot-Watt University in Edinburgh supervised the construction of a GALAXY Mark 2 machine which differed from Mark 1 in three ways: the speed of scanning was considerably increased; the former search mode was turned into a coarse measurement mode; and in the fine measurement mode elliptical replaced the former circular scans making it possible to measure faint galaxies as rapidly as stars. It was able to provide in addition to positions, sizes and photographic densities of images also their ellipticities and orientations at rates of 900 stars and 300 galaxies per hour. The new machine which came into actual operation in 1975 was aptly called 'COSMOS' from its ability to measure '*Co*ordinates, *S*izes, *M*agnitudes, *O*rientations and *S*hapes' of images on Schmidt photographs. A year later COSMOS became a national facility operated like UKIRT and UKST by Edinburgh staff for the benefit of all British astronomers. These were major new responsibilities for the Observatory, and when in 1974 the Royal Greenwich Observatory was given the responsibility of operating the projected Northern Hemisphere Observatory – whose site had been determined by the work of Edinburgh astronomers – the two Royal Observatories had become fully engaged in work for the whole British astronomical community.

14

ASTROPHYSICAL RESEARCHES

FROM THE late 1950s the principal astronomical interests of the Observatory were in the fields of the formation and evolution of stars and of the nature of the interstellar medium. Observationally, this entailed the study of stars and clusters of stars at different evolutionary stages and the spectroscopic analysis of the effect on starlight of material in the intervening interstellar space. The work had to include also theoretical investigations aiming at the explanation of the nature of the interstellar medium and of the processes of star formation.

When this work began, the question of the precise structure of our Milky Way Galaxy was very much under discussion. It had become evident that its spiral arms were the regions where young stars and nebulosities abounded and where presumably conditions were particularly suitable for the formation of stars out of the interstellar medium. It was also apparent that stars tended to form in groups or clusters and that therefore the study of distances and ages of star clusters was a promising field of research.

Systematic observations of star clusters with the Schmidt telescope on Blackford Hill were started by Dr Reddish in the 1950s. They were supplemented by investigations by Dr M.J. Smyth, Mr A. McLachlan and others of star clusters in the southern hemisphere using material from the Armagh-Dunsink-Harvard telescope. The first substantial work on clusters involving Edinburgh observations was a detailed study by Dr Reddish and associates of the magnitudes and colours of 3000 stars in the well-known cluster of the Pleiades which led to a much improved picture of the state of evolution of its faintest members. This was followed by an investigation by Dr N. M. Pratt of the clusters H and Chi Persei where a notable attempt was made to study the polarisation of starlight photographically. A dramatic piece of work was the discovery by Mr L. C. Lawrence and Dr Reddish of the populous nature of the group of stars, called the Cygnus OB2 association, heavily obscured behind dark interstellar dust clouds, but revealed in the Edinburgh work as consisting of thousands of young intrinsically blue stars, comparable in size with large globular star clusters. Globular clusters in the Galaxy are generally made up of stars in late stages of evolution and it was of considerable interest to find that such large structures could still be formed in relatively recent times. The Cygnus association is the only object of its kind ever recorded in the Galaxy, though the two neighbouring galaxies, the Magellanic Clouds, contain a number of young blue globular clusters.

Work on the connection between young star clusters and the interstellar medium went hand in hand with more direct studies of the dust component of that medium. These were carried out through spectroscopic observations of the effect of that component on the intensity distribution in the spectra of distant stars lying behind clouds of obscuring matter. The actual observations were made with the Schmidt telescopes of the Observatory first on Blackford Hill and later at Monte Porzio. The telescopes had been fitted with objective prisms and the photometric calibration of the plates had been achieved through the 'prism-crossed-grating method' using coarse gratings made of threads of nylon and constructed in the Observatory workshops.

These observations by Dr K. Nandy led to a series of important papers in the *Observatory Publications* which gave the effect of the interstellar obscuration or reddening on stars in different regions of the Galaxy indicating the general nature of the small dust grains responsible for the effect.

The first more detailed theoretical conclusions drawn from the Edinburgh observations and published by Drs K. Nandy and N. C. Wickramasinghe in the *Observatory Publications* in 1965 pointed to a picture of the dust grains as consisting largely of graphite covered with ice mantles. In the course of time that picture became increasingly complex and in the early 1970s when optical observations had been supplemented by the first satellite observations in the ultraviolet as well as by other observations in the infrared, interstellar dust grains appeared to consist of mixtures of graphite, iron, silicate and quartz particles. An important summary of the work on interstellar grains appeared in 1972 in *Reports on Progress in Physics*.

Pursuing the link between clouds of interstellar dust and young stars, a search was made on Mount Palomar Schmidt photographs for dark 'globules', spheres of obscuring material within which stars were presumed to be in a state of actual formation, and a catalogue of such globules was published by Miss M. E. Sim in 1968 in the *Observatory Publications*. The work was extended to a scrutiny of young objects of all kinds, and an apparent deficiency of such objects in certain regions of the Galaxy was shown to be capable of interpretation in terms of distinct variations in obscuration by different grains in the intervening spaces.

A further branch of this field of study concerned the investigation of individual dust-embedded stars and star clusters which appeared to show that the youngest stars are generally the most obscured, indicating that they are still enclosed in their original dusty cocoons. These highly obscured objects were the topic of several publications including those of Dr W. Samson, a Research Student in the University Department who used the computer-controlled iris plate photometer to deal with the actual measurements of the extensive material of photographs of star clusters.

Investigations of a related nature of young clusters including the stellar association in Orion were pursued in a search for the very youngest members of the groups, suspected by their colours to be still in the process of gravitational contraction out of their diffuse protostellar phase. These contracting stars, among them the so-called T Tauri stars, are frequently variable stars which emit much infrared radiation on account of the dust shells which surround them. Identification of such stars by means of their characteristics was a task well suited to the programmable plate photometer which could be made to record the colours and light variations of a list of stars observed on a large number of photographic plates. The relevant computer programs for the Orion work were prepared by Miss M. E. Sim.

When the GALAXY measuring machine came into operation in 1969 really large numbers of stars could be tackled and traced through the many thousands of images on photographic plates taken through different colour filters. In this way stars could be segregated according to distance leading to the mapping of the nearer spiral arms of our Galaxy. Such work was performed by Drs R. J. Dodd and W. Sherwood.

An unusual contribution to the study of star formation in the interstellar medium was made by experiments carried out at the UK Atomic Energy Laboratory, Culham, by T. J. Lee, who was then on the staff of Culham but later joined the staff of the Observatory in Edinburgh. These experiments which Dr Lee pursued in collaboration with Dr Reddish, simulated conditions in interstellar clouds composed of molecules of gas and solid grains, and recorded the condensation in such conditions of molecular hydrogen on the surfaces of grains of various kinds.

The perplexing problem of the actual mechanisms of star formation was one which occupied Dr Reddish over many years, and which he tackled from the theoretical as well as from the observational and experimental angles. Calculations as to how, for example, fragmentation of large interstellar molecular clouds would occur, and whether there would be a preference for the larger mass fragments to be found in the outer parts of the resulting aggregates of stars, formed the basis of a number of papers.

Yet another aspect of studies of the interstellar medium was the refinement of earlier observations of interstellar absorption bands in the spectra of distant stars. Edinburgh interest in these features dates back to the spectrophotometric programme on early-type stars of Greaves, Baker and Wilson. That work had shown that the main absorption band falls in the blue part of the spectrum at 4430 Å and is a broad shallow depression in the spectrum caused by an unknown interstellar agency. That early Edinburgh work had also uncovered the existence of a number of other equally inexplicable interstellar bands at other wavelengths. In the more recent work profiles for up to a dozen such bands were derived from spectra obtained with both the 36-inch telescope on Blackford Hill and the 98-inch Isaac Newton Telescope of the Royal Greenwich Observatory. Some of the Edinburgh spectrograms were secured with a Spectracon image tube attached to the spectrograph of the 36-inch telescope. The analysis of these various observations led to the conclusion that these absorption bands may well be explained as due to impurities in the silicate or graphite grains of the interstellar medium. Most of the observational and theoretical work for this programme was carried out by joint teams with Dr Nandy and Mr Seddon from the Royal Observatory and Drs P. W. J. L. Brand, M. T. Brück and G. E. Bromage from the University Department. Image tube observations of the interstellar bands were also made with the 74-inch telescope of the Radcliffe Observatory, Pretoria, by Dr W. Zealey leading, as in the case of Dr Bromage, to the award of PhD degrees.

The study of the nature of the interstellar medium was further broadened when Dr R. Wolstencroft came to Edinburgh, first as a Lecturer in the University Department and later as a member of the Observatory staff. His arrival led to an extension of the spectrophotometric to spectropolarimetric investigations of the interstellar bands which were carried out at both the Edinburgh and Cambridge Observatories.

In 1970 an effort was made to extend this work on the interstellar medium in our Galaxy to a study of interstellar matter in external galaxies of which the nearest ones, the Magellanic Clouds were chosen. The necessary spectra were obtained by Dr A. D. Thackeray with the 74-inch telescope at the Radcliffe Observatory and their analysis was pursued by a joint Edinburgh-Radcliffe team. Drs M. T. Brück, K. Nandy and Mr Lawrence in Edinburgh and Dr A. D. Thackeray at Radcliffe succeeded in deriving for the first time the effect of interstellar matter in the Magellanic Clouds on the light of stars in those Clouds. Their results showed that diffuse matter in the Magellanic Clouds has the same effect on starlight as the interstellar matter in our Galaxy and that the interstellar Cloud material is likely to have the same constitution as the equivalent material in our own Galaxy.

When the first successful Edinburgh rocket observations were analysed by Dr J. Campbell they provided, apart from data on stellar spectra in the ultraviolet and stellar radiation fluxes, a certain amount of information on interstellar absorption. However, the next major advance in the Observatory's studies of the interstellar medium was naturally bound up with the launch in 1972 of the TD1 satellite which made it possible to extend the investigation of the interstellar absorption law from the visual to the far ultraviolet spectral range out to about 1400 Ångstrom. The joint Edinburgh-Liège team concerned with the satellite succeeded in collecting observations of the spectra of some hundred stars in three regions of the Milky Way, corresponding to the Centre and Anticentre of the Galaxy and to an area in Cygnus. Their principal discovery was of a strong interstellar absorption band at a wavelength near 2200 Ångstrom similar to, but much wider than the best-known interstellar band in the blue at 4430 Ångstroms. The ultraviolet observations combined with the earlier observations in the visual part of the spectrum resulted in a 'definitive' interstellar absorption curve for the effect of dust on the light of stars in our Galaxy. This curve, ranging all the way from the infrared to the ultraviolet, was built up by the work of the combined Edinburgh-Liège-London team of Drs Nandy, Thompson, Jamar, Monfils and Wilson. Models for the dust grains responsible for the interstellar absorption have been proposed by Drs Nandy and Morgan.

Some of the spectroscopic work on the interstellar medium and, in particular, the analysis of image tube spectra had been greatly assisted by the methods of automated spectrophotometry which had been developed by Dr G. I. Thompson in the 1960s. This work had been originally planned as a replacement of the methods of spectrophotometry initiated by Professor Greaves and Dr Baker. The new programme while preserving the special characteristics of the 'Edinburgh method' – the sampling of spectrograms at large numbers of points and the combination of data from several similar stars – was to ensure greatly increased speed of operation through the introduction of automatic methods of measurement and computerised procedures of reduction.

The purpose of the new spectrophotometric programme was the study of chemical abundances in stellar atmospheres and of their differences in stars of different stellar populations. This meant in practice the study of line intensities in the spectra of medium-type stars in which absorption lines are far more numerous than in the relatively simple spectra of Greaves' original programme. The spectrograms for this programme were obtained with a powerful new grating spectrograph attached to the 36-inch telescope. They were run through what were then the Observatory's newly digitised Hilger and Joyce-Loebl recording microphotometers.

The application of the complex reduction programme of the observational data – which was published by Dr Thompson in the *Observatory Publications* – was held up by the lack of computing facilities available to the Observatory at the time. The work had to depend on the use of the EDSAC computer in Cambridge, Atlas in Manchester, and an ICS-machine in Edinburgh. The data could at last be handled efficiently when the Elliott 4130 computer had been installed in the Observatory. In its final form the programme made it possible to deal with the analysis of virtually unlimited ranges of spectrum at remarkably high speed. Spectrum lines were found completely automatically, blending of neighbouring lines was allowed for and the background continuous spectrum was determined by a method which eliminated all possible systematic errors. The result of this major piece of work which was published by Dr Thompson in 1972 in the *Observatory Publications*, included a list of intensities of 600 lines in some 80 spectra of 21 medium-type stars. Their analysis indicated the absence of any anomalies in chemical abundances such as had been quoted for the same stars in the earlier literature.

It was very unfortunate that this exceptionally promising spectrophotometric work had to be abandoned in favour of the reduction of the data collected by the TD1 satellite for which Dr Thompson's experience proved invaluable. Both he and Dr Napier transferred their efforts to the satellite work with which Dr Thompson stayed throughout the lifetime of TD1 in company with Dr Nandy and the Liège and London astronomers. The speedy and efficient analysis of the large amount of observational data on the radiation of stars in the ultraviolet which was collected in the course of the sky scan of the TD1 satellite, became the responsibility of the Edinburgh astronomers in the Edinburgh-Liège-London team. In the course of this work Drs Humphries, Nandy and Thompson worked on the ultraviolet energy distribution in young blue stars of different luminosities before the joint team published its final catalogue of ultraviolet data for some 7000 stars.

As the scientific staff of the Observatory increased over the years new fields of work were introduced in addition to those of star formation and interstellar studies which had formed the main theme of the Observatory's research. Much of the new work was published in various scientific journals rather than in the *Observatory Publications*. These papers – some two hundred of them between 1957 and 1975 – were reprinted and issued as *Observatory Communications*.

Among these investigations was the observational and theoretical work by Drs Napier, Smyth and Stobie on variable stars with more than one period of light variation and on the reflection effect in close double stars. Drs B. N. G. Guthrie and J. G. Ireland discussed in a number of papers the effect of stellar rotation on line intensities in the spectra of early-type stars.

The collaboration between the Edinburgh and Rome observatories resulted in joint papers by Dr P. Smirglio of Rome and Dr K. Nandy which dealt with the distribution of late-type stars in various directions of the Galaxy. The observational material for this programme came from both the Edinburgh station at Monte Porzio and the Rome station at Campo Imperatore on the Gran Sasso d'Italia.

By 1970 a group led by Dr M. J. Smyth had started work on infrared spectroscopy using the 98-inch Isaac Newton Telescope of the Royal Greenwich Observatory, the 74-inch telescope at the Radcliffe Observatory in Pretoria and the 60-inch flux collector on Tenerife. Dr Smyth with Messrs Cork, Harris and Wallace obtained infrared spectra in the range from 1 to 2.5 microns in which they made detailed studies of the carbon monoxide bands. Another investigation was that of the infrared light curve of the eclipsing binary Algol by Drs Smyth, Napier and Dow. These infrared observations marked the early stages of what was to become one of the major fields of research at Edinburgh after the completion in 1979 of the UK Infrared Telescope.

The existence of a powerful computer at the Observatory opened up naturally the possibility of dealing with theoretical problems which involve complex numerical calculations. Examples of such work are the papers by Drs Dodd and Napier which are concerned with the question of the origin of the asteroids and with the mass distribution of protoplanets in the solar system which were studied by means of Monte Carlo simulations of collision processes.

The Observatory's work was extended in an important new direction when Dr S. V. M. Clube joined the senior staff in 1972. In contrast to the purely astrophysical nature of most of the Observatory's work Dr Clube's field of specialisation was the study of the motions of stars and the effect of new interpretations of such motions on ideas about the dimensions and the rotation of the Galaxy. This sphere of activity had not been pursued in Edinburgh since the days of Sir Frank Dyson.

Nearly all the Observatory's work described so far was concerned with problems concerning the constituents, stars and interstellar matter, of our own Galaxy, the simple reason being that until well into the 1970s the Observatory had to rely almost entirely on its own observational resources which were modest in terms of the size of its telescopes. The first extragalactic investigation of significance was a survey published in 1973 of the distribution of very young stars in the Large Magellanic Cloud, the nearest external galaxy, by Drs Dodd, Nandy and Wolstencroft. They used photographic plates taken with the American Curtis Schmidt Telescope at Cerro Tololo, Chile, which they measured with the GALAXY automatic measuring machine.

The Observatory had not long to wait, however, before it could launch a substantial extragalactic programme of its own. At the end of 1973 the first photographs reached Edinburgh from the new UK Schmidt Telescope in Australia. The primary purpose of UKST was to photograph the entire southern sky, but additional photographs surplus to the survey and photographs of less than perfect quality were to be made available for research purposes. These were shipped to Edinburgh and stored in the Observatory for the use of the British astronomical community. The responsibility for

The Virgo Cluster of galaxies. Photograph taken with the 48-inch UKST.

the survey of the southern sky as well as that of organising the distribution of photographs for research purposes was in the hands of Dr R.C.Cannon who had joined the Schmidt team in 1973 and was at the same time engaged in research on globular star clusters.

No time was lost in Edinburgh in putting the magnificent new observational material to use. By that time the COSMOS measuring machine was fully operational and computer software for the analysis of faint images on Schmidt photographs had been prepared by the COSMOS team (Dr N.M. Pratt and Mr R.Martin) and the members of the Computer Applications Service of Heriot-Watt University (Messrs L.W.G.Alexander, G.S.Walker and P.R.Williams). The combination of the splendid material of plates with the existence of a fast machine capable of analysing them represented the culmination of what the Observatory had been working towards for some fifteen years.

The first use of the Schmidt photographs which reach objects of 23rd magnitude, was devoted to a survey by Drs Dodd, Morgan, Nandy, Reddish and Mr Seddon of the positions, sizes, shapes and orientations of the images of 3000 faint galaxies and to a discussion of the evidence for their clustering in space. An important investigation by Dr H.T.MacGillivray and various members of the Observatory and Heriot-Watt University staff led to the development of a method which allows the computer to distinguish automatically between the images of galaxies and those of stars on photographs analysed by COSMOS. Results have shown that such a separation can be achieved in at least 90 per cent of the cases. A search for very faint blue objects on deep Schmidt photographs by Drs Reddish and M.

Hawkins revealed a number of objects identified in all probability as quasars. This exercise was a test of the capability of COSMOS to identify objects with unusual characteristics amongst hundreds of thousands of other images.

Among the earliest Schmidt photographs dispatched to Edinburgh was a large set of plates taken through various colour filters of the nearby galaxy, the Small Magellanic Cloud. This material led to an investigation by Dr M. T. Brück by methods used earlier in the Observatory's work on galactic star clusters of the spatial distribution of some 300 star clusters in the Magellanic Cloud. This work was continued and used as a thesis for an Edinburgh doctorate by Dr Mary Kontizas of the University of Athens. Another specialised use of the COSMOS machine was the mapping of extended gas and dust structures in fields of the Milky Way. A programme aiming at an interpretation of certain types of such structures in terms of energetic processes providing clues to mechanisms of star formation was started in 1975 by Drs Brand and Zealey as another promising new field of work.

The programmes mentioned here are examples of the work carried out at the Observatory with the new Schmidt photographs within the two years which elapsed between the arrival in Edinburgh of the first batch of plates and my own retirement from the Observatory. I believe that they give an indication of how well the Observatory was prepared for the exploitation of the new observational material. Many other programmes had been started or planned by 1975, including joint projects with groups in other Universities. Shortly afterwards the COSMOS machine was declared like the Schmidt telescope itself a national facility, so that both the material and the means of collecting and analysing it were from that time onwards formally at the disposal of all interested astronomers in the United Kingdom.

15

THE CRAWFORD COLLECTION

THE GIFT of the 26th Earl of Crawford to the nation in 1888 to which the Royal Observatory owed its survival included as its most valuable part the historical collection of early books and manuscripts some of them dating back to the thirteenth century. Though many generations of the Crawford family had been collectors and an earlier library known as the old Balcarres collection was reported to contain the finest scientific collection in 17th-century Scotland, the particular historic library in the Crawford gift was for the most part a more recent collection and remarkable in that it was put together in a short space of years by one man and not by generations of librarians.

The father of the 26th Earl had commented that the 'scientific section' of his library was weak and in 1870 empowered his son, then Lord Lindsay, to rectify this want. Within about ten years Lord Lindsay could say that he

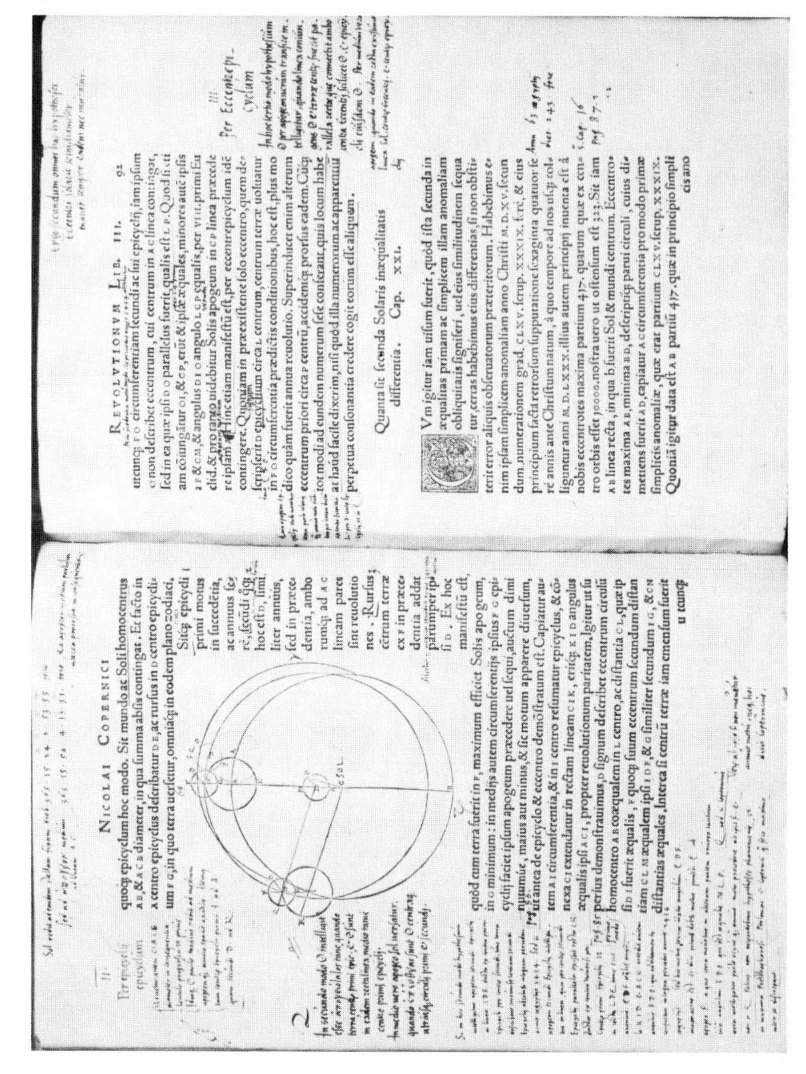

A page from Copernicus' *De Revolutionibus Orbium Coelestium*, with annotations by Erasmus Reinhold.

had now 11,000 books forming one of the great collections in the world. When setting out to find the books he used as his guide the catalogue of the Imperial Observatory at Pulkovo near St Petersburg, at that time the greatest scientific library anywhere, whose Director, Otto Struve, gave him every possible advice.

The nucleus of the collection was formed from the great library of Charles Babbage, at one time Lucasian Professor of Mathematics at Cambridge and the inventor of computers, which was sold after his death in 1871. This collection alone contained more than 2500 items. In addition, rare books were systematically purchased from many parts of the Continent of Europe and some were added from the library of the Crawford family at Haigh Hall near Wigan in Lancashire.

On the occasion of the formal Opening in November 1972 of the new 'Crawford Room' at the Royal Observatory where this unique historical collection is now housed, the 28th Earl of Crawford, grandson of the original collector and donor said of his grandfather: 'It was fundamental to his point of view that the history of science was an important subject in itself, a subject in which scientists could and should take an interest, and it was because of this that he gave not only his own working library, with all the necessary periodicals, and up-to-date books, belonging to his own observatory, but also the historical section which forms the glory of the Library'.

The Library contains the first editions of nearly every book important in the history of astronomy and related fields. It is especially rich in early literature on comets and there are many treatises on astrology.

The pre-Copernican period of astronomy is represented by the earliest printed edition of 1478 of Sacrobosco's (John Holywood's) *Sphaera Mundi*; by a near perfect copy of Apianus' *Astronomicum Caesareum* of 1540, with working planispheres; and by many other volumes.

There is a beautiful copy of Copernicus' *De Revolutionibus Orbium Coelestium* of 1543 with annotations by Erasmus Reinhold, who was Professor of Mathematics and Astronomy in Martin Luther's University at Wittenberg from 1536 to 1553. There is a copy of Reinhold's *Tabulae Prutenicae* of the motions of celestial bodies calculated on the basis of Copernicus' *De Revolutionibus*. The Collection also contains a copy of the first edition of 1627 of Kepler's *Tabulae Rudolphinae*, which superseded Reinhold's and remained the standard astronomical tables for more than a century.

Amongst the many Kepleriana in the Collection mention may be made of Kepler's pamphlets on the new star which appeared in 1604, and of his famous book *Harmonices Mundi*, which was published in 1619. The Collection's copy holds within its covers an autograph sheet by Kepler, being part of a contract and specification for a hydraulic engine. There is also a copy of Kepler's *Astronomia Nova*, printed in 1609 and containing Kepler's well-known first two laws which he established from Tycho Brahe's observations of the motion of the planet Mars.

Tycho Brahe's own publications are very well represented, starting from his observation of the Nova of 1572 to the description of his instruments in *Astronomiae Instauratae Mechanica* of 1602.

QVADRANS MVRALIS
SIVE TICHONICUS.

HARMONICIS LIB. III. 61

CAPVT XIII.
Quid ſit Cantus naturaliter Concinnus & aptus.

Nihil dicemus de ſtridulo illo more canendi, quo ſolent uti Turcæ & Vngari pro claſſico ſuo: brutorum potius animantum voces inconditas, quam humanam Naturam imitati.

Videtur omninò primus author hauſiſſe melodiam hujuſmodi inconditam ab inſtrumento minᵘ aptè conformato, eamq; conſuetudine diuturnâ, cum ipſiᵘ inſtrumenti facturâ, tranſmiſiſſe ad poſteros, totâq; gentem. Interfui Pragæ precibᵘ, quas Legati Turcici ſacerdos horis ſtatis ingeniculatus, terramq; fronte crebrò feriens, decantare ſolebat: apparuit facile, ipſum ex diſciplina canere, exercitationemq; & promptitudinem labore comparaſſe, nihil enim hæſitavit: at intervallis uſus eſt miris, inſolitis, conciſis, abhorrentibus, ut nemo proprio naturæ ductu & ex ſeipſo ultrò ſimile quid conſtanter unquam meditari poſſe videatur. Conabor aliqd proximũ illi per noſtras notas Muſicas exprimere.

Concinnus igitur & humanarum aurium judicio aptus cantus eſt, qui exorſus à certo quodam ſono; ab eo per intervalla concinna tendit ad ſonos conſonos & primo illi, & plerumq; etiam inter ſe mutuò; diſſona curſim pervolitans intervalla, in conſonis verò immorans, ſeu menſurâ temporis, Syllabarumq; longitudine, ſeu crebro ad illos reditu, veluti duarum vocum inter ſe conſonantiam affectans; unicæ vocis traductione à loco uno Syſtematis ad alium. Exemplum.

Hic ſonus initialis eſt in clavj G, cum quâ in cantu molli concordant b. c. d. g. Excurrit igitur Cantus (primùm flexu deorſum facto) ad clavem c. conſonam, & tranſilit planè diſſonum locum A; fuiſſet autem idem, ſi attigiſſet ipſum, ſed brevi mora; tota verò ſeries reliquę potiſſimùm in locis b. d. g. intonat, ſceletó octavæ tale exprimens, in d. creberrimè rediens, pòſt in b. in ſuperius verò g. ſe interdum efferens, in hæc omnia loca ſignanter: non ſic in a. vel in f. loca primo diſſona: tandemq; redit ad G. ibiq; finit.

Circa traditam cantus definitionem multa veniunt nobis annotanda.

I. Partes Cantus, ex quibus vel omnibus vel aliquibus conſtat omnis Cantus, Euclides nominat has quatuor, Ἀγωγὴν, Τονὴν, Πεττείαν, Πλο-

H 3

Left: Tycho Brahe's Mural Quadrant from his *Astronomiae Instauratae Mechanica*.
Right: From Kepler's *Harmonices Mundi*.

RECENS HABITAE. 20

di medium iam inter Iouem, & orientalem Stellam locum exquiſitè occupantem, ita vt talis fuerit confi-

Ori. * * O * * Occ.

guratio. Stella inſuper nouiſſimè conſpecta admodum exigua fuit; veruntamen hora ſexta reliquis magnitudine ferè fuit æqualis.

Die vigeſima hora 1. min: 15. conſtitutio conſimilis viſa eſt. Aderant tres Stellulæ adeo exiguæ, vt vix

Ori. * O * * Occ.

percipi poſſent; à Ioue, & inter ſe non magis diſtabant minuto vno: incertus eram nunquid ex occidente duæ, an tres adeſſent Stellulæ. Circa horam ſextam hoc pacto erant diſpoſitæ. Orientalis enim à Ioue

Ori. * O** Occ.

duplo magis aberat quam antea, nempe min: 2. media occidentalis à Ioue diſtabat min: 0. ſec: 40. ab occidentaliori vero min: 0. ſec: 20. Tandem hora ſeptima tres ex occidente viſæ fuerunt Stellulæ. Ioui proxima abe-

Ori. * O*.* Occ.

rat ab eo min: 0. ſec: 20. inter hanc & occidentaliorem interuallū erat minutorum ſecundorum 40. inter has vero alia ſpectabatur paululum ad meridiem deflectēs; ab

PHILOSOPHIÆ
NATURALIS
PRINCIPIA
MATHEMATICA.

Autore *J S. NEWTON, Trin. Coll. Cantab. Soc.* Matheſeos Profeſſore *Lucaſiano*, & Societatis Regalis Sodali.

IMPRIMATUR.
S. PEPYS, *Reg. Soc.* PRÆSES.
Julii 5. 1686.

LONDINI,

Juſſu *Societatis Regiæ* ac Typis *Joſephi Streater.* Proſtat apud plures Bibliopolas. *Anno* MDCLXXXVII.

Left: The observations of Jupiter's satellites from Galileo's *Siderius Nuncius*. *Right:* Frontispiece of the first edition of Newton's *Principia*.

MIRIFICI

Logarithmorum
Canonis deſcriptio,

Ejuſque uſus, in utraque
Trigonometria; ut etiam in
omni Logiſtica Mathematica,
Ampliſsimi, Facillimi, &
expeditiſsimi explicatio.

Authore ac Inventore,
IOANNE NEPERO,
Barone Merchiſtonii,
&c. Scoto.

EDINBURGI,
Ex officinâ ANDREÆ HART
Bibliopolæ, CIↃ. IↃC. XIV.

Frontispiece of Napier's first treatise on logarithms.

The Collection's many books by Galileo include a fine copy of his *Sidereus Nuncius* of 1610, with his observations of Jupiter's satellites amongst others; a copy of *Istoria e Dimostrazioni* of 1613, with his earliest drawings of sunspots; and *Il Saggiatore* of 1623, a well-known piece of his controversial writing.

The Collection's editions of Isaac Newton's *Principia Mathematica* include the first edition of 1687, with an imprimatur by S. Pepys, then President of the Royal Society, and the third edition of 1726, with Newton's portrait by Vanderbank.

Amongst mathematical treatises in the Collection are some thirty editions of Euclid. These include the first Latin edition (Venice, 1482), the very rare Paris edition of 1516, the first Greek edition (Basle 1533), the first English edition (London 1570), and the first Arabic edition (Rome 1594).

Of particular interest to Scottish readers are the Collection's treatises on logarithms by Baron John Napier of Merchiston. They include his announcement of their discovery and his first logarithm table which he published in 1614 under the title *Merifici Logarithmorum Canonis Descriptio*.

In the early years on Blackford Hill, the Crawford Collection and indeed the entire Observatory library was looked after by the Astronomer Royal of the day. Only in 1930 does the list of Observatory staff include a lady giving 'general assistance in the library as well as performing computing and secretarial duties'. It was well into the 1950s before Professor Greaves succeeded in getting additional assistance for this hard-pressed secretary and in getting approval for the appointment of two clerical officers.

By 1960 books and periodicals and the number of users of the Observatory

The Earl of Crawford and Balcarres at the opening in 1972 of the Crawford Room of the Observatory Library. Right to left are Lord Flowers, Chairman of the Science Research Council, Lord Crawford, Lord Swann, Principal of Edinburgh University, and the author.

library had increased to such an extent that it became imperative to have a trained librarian on the staff, and in 1961 the Observatory acquired its first full-time professional librarian, Mr D. A. Kemp. In the following years Mr Kemp started a much-needed reorganisation of the library and at the same time Mrs Mary F. I. Smyth became assistant librarian with the special charge of looking after the contents of the Crawford Collection as part of the general reorganisation.

Until this time the most valuable items in the Crawford Collection were kept locked away in a strong room and there was no easy access to them by research workers. The Catalogue of the Collection started by Lord Lindsay at Dunecht had been completed by Professor Copeland and published by him in 1890 after the Collection had been transferred to the Royal Observatory on Blackford Hill. By 1970 there was need for a revised catalogue which would also include items from the former Calton Hill Observatory and old books acquired by Professor R. A. Sampson.

The entire Collection was checked and re-catalogued by computer and at the same time a special 'Crawford Room' was created within the Observatory as part of its sesqui-centenary celebrations in 1972 where the most valuable portions of the benefaction could be held in conditions of perfect safety under complete temperature and humidity control. The work of re-cataloguing the Collection by computer was carried out by Dr M. J. Smyth and Mrs Mary F. I. Smyth with the assistance of the Observatory's librarian, Mr A. R. Macdonald.

The formal opening of the new Crawford Room in November 1972 by the late Earl of Crawford took place in the presence of the Chairman of the Science Research Council, Sir Brian (now Lord) Flowers, the Principal of the University, Sir Michael (now Lord) Swann and Sir Douglas Haddow of the Scottish Office. On that occasion a detailed account of the history of the Crawford Collection and its contents was given by Professor Eric Forbes, Professor of the History of Science in Edinburgh University. The proceedings were published in Volume 9 of the *Observatory Publications* which also contains a description of the computer cataloguing of the Collection by Dr M. J. and Mrs Mary F. I. Smyth.

The Crawford Collection has been used by many historians of science, including Professor Doris Hellman of New York, Professor Owen Gingerich of Harvard and Professor Stanley Jaki of Princeton. Dr Giovanna Grassi Conti of the University of Rome spent several months at the Observatory in pursuit of her work on a Catalogue of 15th and 16th century astronomical publications in the libraries of European observatories.

With the new Crawford Room, with a computerised index and with accessible reading facilities Lord Crawford's great collection is now fully available to scholars from all over the world.

16

ASTRONOMY TEACHING IN THE UNIVERSITY

AS HAS been explained in an earlier chapter astronomy came into being in the University of Edinburgh in 1786 with the establishment of a Chair of Practical Astronomy. However, serious teaching of the subject started only a hundred years later. The first holder of the Chair never lectured and took little or no interest in University affairs while his successor, Thomas Henderson, devoted all his energies to his intense observational work at the Calton Hill Observatory and, being the first Astronomer Royal for Scotland, did not consider himself under an obligation to institute formal teaching courses.

It was only in 1850 that the first course in astronomy was offered when Charles Piazzi Smyth began his lectures four years after his appointment and a full 64 years after the foundation of the chair. Even then astronomy teaching did not take root. In his later years in office Piazzi Smyth became disillusioned with the Professorial side of his post and confined his teaching to occasional instructions. Sir Alexander Grant in his *Story of the University of Edinburgh* could well write in 1884 – while Professor Piazzi Smyth was still in office – that 'the Chair of Practical Astronomy which has now been in existence nearly a hundred years has contributed next to nothing to the educational resources of the University'.

It was very unfortunate that this could be said in the case of Piazzi Smyth who was in many ways a remarkable expositor of astronomy in its widest sense. As already recounted, he made a major effort over a period of four or five years to establish a worthwhile course of instruction. His lack of success had as much to do with the unsatisfactory definition of the subject of 'Practical Astronomy' as with his own highly individual and impatient personality. There was also little material support – a mere £10 was granted to purchase equipment for teaching – and Piazzi Smyth had to rely on instruments borrowed from his father's private observatory before he could launch his course.

'Practical Astronomy' though still remaining in the Calendar, effectively lapsed in the University in Piazzi Smyth's later years, and in his Annual Reports to his Board of Visitors we find him referring to the title of Professor as a purely honorary one, and regarding himself as having no duties other than those of the Astronomer Royal for Scotland.

Fortunately, Piazzi Smyth's much publicised quarrel with the University Court over his Professorial duties in no way impaired his friendship with many of his University colleagues. Those who like Professor Tait, Professor of Natural Philosophy, were members of his Observatory Board of Visitors, always supported him in his efforts to get more recognition for the work of the Calton Hill Observatory.

Though students did not directly benefit from Piazzi Smyth's vast knowledge and experience of advanced techniques in physical astronomy and laboratory spectroscopy many of which he himself had pioneered, the University fully recognised his many contributions to science by making him in 1890 an Honorary Doctor of Laws, a distinction which he was very proud to carry for the rest of his life.

At the Tercentenary Celebrations of the University in 1884 Piazzi Smyth was the formal representative of the Royal University of Palermo, the city with which he had a life-long association and of whose famous Academy he was a long-standing member. For some reason, probably because of his extreme deafness at the time, he delegated the duty of attending the official ceremony to his friend and associate in his spectroscopic work, Professor A. S. Herschel of Durham, the son of Sir John Herschel, who was his guest during the week. Another guest was Sir David Gill, HM Astronomer at the Cape, who represented the University of Cape Town and who was awarded by the University at the same time an Honorary LLD on the recommendation of Piazzi Smyth.

When Copeland became Professor on Piazzi Smyth's retirement he did not feel himself constrained in any way by the archaic duties attached to the Chair at its foundation. He prepared a conventional course which included spherical, positional and orbital astronomy and, in particular, stellar spectroscopy. This was the 'intermediate' class which was intended as the normal course for students of Mathematical and Natural Philosophy. In 1891 he added an advanced class which was given whenever there was sufficient demand for it. This course counted as a half course for final year students in Natural Philosophy and covered what was then called 'theoretical astronomy' which meant the advanced treatment of positional and dynamical astronomy. The level was high and students may have found it hard going; perhaps this is why we find Copeland urging on the University authorities the desirability of a Chair of Applied Mathematics. Unlike

Piazzi Smyth who was obliged to operate in isolation, Copeland through the advanced class maintained close links with Physics. Practical classes held at the Observatory involved in addition to the use of astronomical telescopes and auxiliary instruments, experiments in optics, spectroscopy and terrestrial magnetism. These classes were held at the Calton Hill until the Blackford Hill building was ready, though there is a record of one student being taken up to Blackford Hill with portable instruments at a very early stage in the building operations. When the new Observatory was planned Copeland took special care to include an optical laboratory for undergraduate teaching and provision on the flat roof of the Observatory for erecting portable instruments. Emphasis was placed on spectroscopy which was then the most exciting field of 'physical astronomy'. Copeland's courses were the beginning of serious undergraduate teaching in astronomy and helped to shape such courses for half a century or more. For ten years they were taught singlehandedly by Copeland who delivered 80 lectures per year and held a two-hour practical session each week at the Observatory. It is interesting to note that from 1892 when women students were first admitted into the University, there were always some in Copeland's Astronomy class; in fact the proportion of women students studying astronomy is no higher today than it was then. In the 1890s Copeland also acted as external examiner to Professor Sir Norman Lockyer of the Royal College of Science in London, the famous discoverer of helium in the Sun. In Copeland's last years the advanced course was given by his chief assistant, Dr Halm, who continued to teach it under Professor Dyson until he himself left Edinburgh in 1907.

A special collection of books for the use of undergraduates, the Ramsay Memorial Library, was set up in 1901 in the Observatory after the death of a promising young astronomer, Andrew James Ramsay, who had joined the Calton Hill Observatory as a student in 1892 and who a year later had been taken on as a second assistant by Professor Copeland. Ramsay accompanied Copeland on his solar eclipse expeditions to Norway and India and became the Observatory's official photographer. His untimely death in 1899 after a short illness led to the establishment by friends of the Ramsay Memorial Trust which since 1902 has provided each year a book prize for the best student in the Astronomy class.

Professors Sampson and Greaves continued what came to be called the Astronomy Ordinary course and occasionally gave specialised courses to interested students in the final honours year. Before the first University Lecturer in Astronomy was appointed in 1951 Professor Greaves had the assistance of Mr A. Nisbet of the Department of Mathematical Physics.

The present Astronomy I course, though naturally greatly changed in substance, follows in direct line from Copeland's Intermediate course, and in the 1960s a 'Fourth Year Option in Astronomy' for Physics students which counted as part of their honours Physics degree bore a strong resemblance in its intentions to the 'advanced' course given between 1891 and 1907.

The next significant change in undergraduate teaching occurred as a result of recommendations in the Robbins Report of 1963 in which a major expansion of undergraduate places was foreseen. By that time the Royal

Top: The 20-inch telescope of the University Department erected in 1967.
Bottom: The Edinburgh University undergraduate student laboratory.

Left: King Hussein of Jordan and Princess Muna at the Observatory on the occasion of their State Visit in July 1966. *Right*: President Cevdet Sunay of Turkey being shown a Skylark rocket when he visited the Observatory in November 1967.

Colonel John Glenn, the first American astronaut, meets a young admirer on his visit to the Observatory in 1966.

Observatory had made substantial advances in equipment and facilities and astronomy had become one of the most exciting and attractive fields of science. The opportunity was perfect for the Department of Astronomy to take its share in the student expansion as well as in the training of a new generation of astronomers which would be needed when the hoped-for new major observational facilities became available in the following decade. The new Astrophysics Honours course was planned as basically a Physics course with the astronomical content treated as a particular extension of Physics. Astronomy was now at a stage when the problems of understanding the nature of stars and of the Universe at large were fundamentally problems of physics. The curriculum of the new course was therefore drawn up in collaboration with the Professors of Physics and Mathematical Physics, N. Feather and N. Kemmer. It resulted in a joint course, called 'Astrophysics', for third and fourth year honours students of physics. The course, the first of its kind in a British University, has proved eminently successful. Several of its graduates have distinguished themselves in astronomical research; others have found that their training equipped them for work in other branches of physics, in instrument technology and in computing.

To accommodate the new honours class, with a target of six graduates per year, a large extension was erected in 1967 at the Observatory by the Science Research Council for the use of the University. This extension consists of three large laboratories, darkrooms and offices. The University Department acquired at the same time a large grating spectrograph to be used for laboratory spectroscopy or, when employed in combination with a horizontal telescope and coelostat, for solar work. The Department also acquired a 20-inch reflector specially designed by Messrs Grubb Parsons in Newcastle as a versatile instrument for advanced student projects.

In order to keep step with the new teaching duties the Departmental staff increased from one to four lecturers. At the same time a scheme was instituted whereby senior members of the Royal Observatory staff deliver specialist courses of lectures to the final year honours students thus further strengthening the link between the Royal Observatory and the University. When the new Department of Geophysics was instituted by the University in 1969, Professor A. H. Cook, the first Professor, delivered his first course of lectures in the Department of Astronomy.

Mention might also be made of the many other students who have regularly visited the Observatory with less serious intent. These are the members of the Edinburgh University Astronomical Society, founded in 1958 and still flourishing. The 10-inch astrograph which had given sterling service to the Observatory in older days, was put at the disposal of members of that Society.

A memorable occasion worth recording in the annals of the Department was the conferring in 1970 of an honorary Doctorate of Science on Professor Albrecht Unsöld of the University of Kiel in Germany, one of the outstanding astrophysicists of his generation.

Following the establishment of the new Astrophysics course I had the privilege of serving (1968–70) as Dean of the Faculty of Science, an indication that the once peripheral Department of Astronomy was able to play a definite part in the life of the University.

EPILOGUE

THE HISTORY of the Royal Observatory Edinburgh differs from that of other national astronomical establishments of the 19th century in that for most of its time the chief interest of its astronomers has been in physical rather than positional astronomy. It could well be argued that of its Directors Thomas Henderson was the only true classical astronomer who was concerned with nothing but the systematic observation of the positions of large numbers of stars. It is true that a substantial amount of positional work was carried out by Piazzi Smyth and Copeland, but looking at their activities one gets the distinct impression that they did that work out of a sense of duty and that their real interest was in what Copeland in his praise of Piazzi Smyth calls plainly astrophysics.

In the 19th century spectroscopy was the key to the study of the physics of the stars and the discovery and identification of new features in astronomical spectra and an ever greater refinement of the exploration of the solar spectrum were amongst the main interests of both Piazzi Smyth and Copeland. Then came the pioneering researches of Sampson in the field of spectrophotometry on the distribution of energy in stellar spectra which were continued and extended by Greaves whose work interpreted the photometric and spectrophotometric data in terms of physical parameters of stellar atmospheres. The Observatory's expertise in the procedures of high precision photographic photometry has remained unquestioned since those days.

My own time in Edinburgh saw the introduction into astrophysics of what was then the new technology which was to enhance the efficiency of the analysis of observational data in photometry and spectrophotometry. Some of the necessary technical work was painfully slow and was not everywhere appreciated. However, though some of our achievements have been overtaken by the technological explosion, we can claim that they have led to the development of machines like GALAXY and COSMOS and through them to many valuable researches which otherwise could never have been contemplated.

My own efforts were greatly helped by the wholehearted enthusiasm of the Observatory staff and also by the fact that in my first eight years at least I alone carried the full responsibility for the work of the Observatory and was under no obligation to refer my programmes to a bureaucratic machinery of Committees. In this way, unusual and – for those days – relatively costly plans such as that of the GALAXY measuring machine or that of the creation

The staff of the Royal Observatory Edinburgh and visiting astronomers on 30 September 1975.

of an observing station abroad could be readily discussed and speedily proceeded with.

I used to describe my job as by far the most attractive astronomical post in the United Kingdom. This was because of the close and indeed unique link between the Government appointment and the University Chair of Astronomy. The possibility of teaching astronomy to students and the presence of students in the Royal Observatory who work there happily side by side with Government astronomers has always seemed to me to give its special strength and flavour to astronomy in Edinburgh.

The era of residential observatories is now long a thing of the past. When I first came to Edinburgh all four houses in the Observatory grounds were occupied by Observatory staff. When I left, my own residence had remained the only inhabited one, following the necessary conversion of the others into offices and laboratories; and this too has now become an extension of the Observatory buildings.

The delightful residence and the marvellous views from Blackford Hill which are so charmingly described by Sir Walter Scott in his *Marmion*:

> Yonder the shores of Fife you saw;
> Here Preston Bay and Berwick Law;
> And broad between them rolled,
> The gallant Firth the eye might note,
> Whose islands on its bosom float,
> Like emeralds chased in gold

are not easily forgotten, as testified by the glowing memories of the families of all four of my predecessors who at one time or another visited us in their old home in the Observatory grounds.

My years on Blackford Hill had been extremely happy ones, but by the time the date of my retirement came around, it had become clear that the ever increasing responsibilities of the Royal Observatory for major national and even international facilities in distant lands required the energy and perspective of a younger generation.

In ending this account of my own and earlier times it is appropriate to say *Valete* to the past and to bid *Ave* to those who are now happily and successfully shaping the next chapter in the history of astronomy in Edinburgh.

BIBLIOGRAPHY

Books

Ball, W. Valentine, *Reminiscences and Letters of Sir Robert Ball*, London, 1915.
Brewster, Sir David, *A Treatise on Optics*, London, 1831.
Campbell, Neil and Smellie, R. M. S., *The Royal Society of Edinburgh (1783-1983)*, Edinburgh, 1983.
Clerke, Agnes M., *A Popular History of Astronomy during the Nineteenth Century*, 3rd Edition, London, 1893.
Clerke, Agnes M., *Problems in Astrophysics*, London, 1903.
Clerke, Agnes M., *The System of the Stars*, London, 1905.
Copeland, Ralph, *Catalogue of the Crawford Library*, Edinburgh, 1890.
Crossland, J. Brian, *Victorian Edinburgh*, Letchworth, 1966.
Dreyer, J. L. E. and Turner, H. H. (Editors), *History of the Royal Astronomical Society, 1820-1920*, London, 1923.
Dyson, F. W., *Astronomy*, London, 1910.
Forbes, George, *David Gill, Man and Astronomer*, London, 1916.
Grant, Sir Alexander, *The Story of the University of Edinburgh during its first 300 years*, 2 Volumes, London, 1884.
Herschel, Sir John, *Outlines of Astronomy*, London, 1849.
Horn, D. B., *A Short History of the University of Edinburgh*, Edinburgh, 1967.
Keir, David (Editor), *The Third Statistical Account of Scotland, The City of Edinburgh*, Glasgow, 1966.
Macpherson, H., *Makers of Astronomy*, Oxford, 1933.
Mendelssohn, K., *The Riddle of the Pyramids*, London, 1974.
Smyth, C. Piazzi, *Report on the Teneriffe Astronomical Experiment of 1856*, London and Edinburgh, 1858.
Smyth, C. Piazzi, *Teneriffe, An Astronomer's Experiment: or, Specialties of a Residence above the Clouds*, London, 1858.
Smyth, C. Piazzi, *Three Cities in Russia*, 2 Volumes, London, 1862.
Smyth, C. Piazzi, *Our Inheritance in the Great Pyramid*, London, 1864.
Smyth, C. Piazzi, *Life and Work at the Great Pyramid*, 3 Volumes, Edinburgh, 1867.
Smyth, C. Piazzi, *On the Antiquity of Intellectual Man*, Edinburgh, 1868.
Tompkins, P., *Secrets of the Great Pyramid*, New York, 1971.
Turnbull, H. W., *James Gregory*, London, 1939.
Turner, A. Logan, *History of the University of Edinburgh, 1883-1933*, Edinburgh, 1933.
Warner, B., *Astronomers at the Royal Observatory Cape of Good Hope*, Capetown and Rotterdam, 1979.
Wilson, Margaret, *Ninth Astronomer Royal, The Life of Frank Watson Dyson*, Cambridge, 1951.
Youngson, A. J., *The Making of Classical Edinburgh*, Edinburgh, 1966.

Papers

Astronomical Institution of Edinburgh, Minute Books, 2 Vols. 1811-1847, Royal Observatory Edinburgh Archives.
Astronomical Institution of Edinburgh, Some Remarks on the Present State and Future Prospects of the Observatory, Edinburgh, 1846.
Barkworth, Fanny S. Copeland, Correspondence with H. A. Brück, 1962-68.
Blair, Archibald, Observations on the Superiority of Achromatic Telescopes with Fluid Object Glasses, *Edinb. Journ. Sci.* Vol. IV, 1826.
Blair, Archibald, On the Permanency of Achromatic Telescopes constructed with Fluid Object Glasses, *Edinb. Journ. Sci.* Vol. VII, 1827.
Blair, Robert, Experiments and Observations on the Unequal Refrangibility of Light, *Trans. Roy. Soc. Edinb.* Vol. III, 1794.
Brück, H. A., The Royal Observatory Edinburgh 1822-1972, Edinburgh, 1972.
Bryden, D. J., James Short and his Telescopes. Royal Scottish Museum, Edinburgh, 1968.
Copeland, Ralph, An Account of some recent Astronomical Experiments at High Elevations in the Andes. *Copernicus Journ.* Vol. 3, 1883.
Copeland, Ralph, Lecture Notes, Royal Observatory Edinburgh Archives.
Crawford and Balcarres, Earl of, Address at the Opening of the Crawford Room in the Royal Observatory Edinburgh. *Publications of the Royal Observatory Edinburgh* Vol. 9, 1973.
Forbes, E. G., The Crawford Collection of the Royal Observatory. *Publications of the Royal Observatory Edinburgh* Vol. 9, 1973.
Gill, A. T., Photography at the Great Pyramid in 1865, *Photographic Journal* Vol. 105, No. 4, 1965.
Herschel, A. S., Some Notes on Charles Piazzi Smyth's Work in Spectroscopy. *Nature,* June 14, 1900.
Isles, J. E., Robert Blair. Manuscript in Department of History, University of Edinburgh, 1969.
Nairne, J., The Laws of the Astronomical Institution of Edinburgh, Edinburgh, 1829.
Pratt, N. M. and Reddish, V. C. (Editors). Astronomical Discussions of Problems of the Interstellar Medium and Galactic Structure. *Publications of the Royal Observatory Edinburgh,* Vol. 4, 1964.
Royal Observatory Edinburgh Reports, 1848-1975.
Royal Observatory Edinburgh Archives. Catalogued by Mary F. I. Smyth. Edinburgh, 1981.
Sampson, R. A., The Growth of Ideas in Astronomy, Edinburgh, 1911.
Sampson, R. A., Bibliography of Books exhibited at the Napier Tercentenary Celebrations in July 1914, Edinburgh, 1915.
Sampson, R. A., Science and Reality, London, 1928.
Scotsman, The, Files in Edinburgh Room of Central Public Library, Edinburgh.
Seddon, H. and Smyth, M. J. (Editors), Automation in Optical Astrophysics. Proceedings of Colloquium No. 11 of the International Astronomical Union held in Edinburgh, 1970. *Publications of the Royal Observatory Edinburgh* Vol. 8, 1971.
Smyth, C. Piazzi, Programme of Lectures in Practical Astronomy, Edinburgh, 1853.
Smyth, C. Piazzi, Description of New and Improved Instruments for Navigation and Astronomy, Edinburgh, 1855.
Smyth, C. Piazzi, Lecture Notes and Diaries preserved at the Royal Society of Edinburgh.

Smyth, M. J. and Seddon, H. (Editors), The 150th Anniversary of the Royal Observatory Edinburgh and Infrared Astronomy in the United Kingdom. *Publications of the Royal Observatory Edinburgh* Vol. 9, 1973.
Warner, B., Charles Piazzi Smyth at the Cape of Good Hope, *Sky and Telescope*, January 1980.

Biographical Notes

in Poggendorff, *Biographisch _ Literarisches Handwörterbuch*:
R. Blair, Vol. I, 207, 1863
T. Henderson, Vol. I, 1064, 1863
C. Piazzi Smyth, Vol. III, 1261, 1898
in *Dictionary of Scientific Biography*:
F. W. Dyson, Vol. IV, 269, 1971
T. Henderson, Vol. VI, 263, 1972
J. Lamont, Vol. VII, 607, 1973
R. A. Sampson, Vol. XII, 95, 1975
C. Piazzi Smyth, Vol. XII, 498, 1975

Obituaries

in *Monthly Notices of the Royal Astronomical Society*:
Sir Thomas Makdougall Brisbane, Vol. 21, 98, 1861
R. Copeland, Vol. 66, 164, 1906 (by J. L. E. Dreyer)
Sir Frank W. Dyson, Vol. 100, 238, 1940 (by J. Jackson)
W. M. H. Greaves, Vol. 116, 145, 1956 (by J. Jackson)
J. Halm, Vol. 105, 92, 1945 (by Sir Harold Spencer Jones)
T. Henderson, Vol. 6, 151, 1845
Sir John Herschel, Vol. 32, 122, 1872
J. Lamont, Vol. 40, 208, 11880
James Ludovic Lindsay, 26th Earl of Crawford, Vol. 74, 271, 1914
R. A. Sampson, Vol. 100, 258, 1940 (by W. M. H. Greaves)
C. Piazzi Smyth, Vol. 61, 189, 1901 (by R. Copeland)
in *Obituary Notices of Fellows of the Royal Society*:
F. W. Dyson, Vol. 3, 159, 1940 (by Sir Arthur Eddington)
R. A. Sampson, Vol. 3, 221, 1940 (by Sir Edmund Whittaker)
in *Biographical Memoirs of Fellows of the Royal Society*:
W. M. H. Greaves, Vol. 2, 129, 1956 (by R. O. Redman)
in *Transactions of the Royal Society of Edinburgh*:
Sir Thomas Makdougall Brisbane, Vol. 22, 589, 1861
in *Proceedings of the Royal Society of Edinburgh*:
F. W. Dyson, Vol. 59, 265, 1940 (by W. M. H. Greaves)
T. Henderson, Vol. 2, 35, 1845 (by P. Kelland)
J. Lamont, Vol. 10, 358, 1880 (by A. Buchan)
R. A. Sampson, Vol. 60, 405, 1941 (by W. M. H. Greaves)
C. Piazzi Smyth, Vol. 23, 48, 1900
in *Yearbooks of the Royal Society of Edinburgh*:
E. A. Baker, 1981, p.5 (by H. A. Brück)
H. E. Butler, 1979, p.15 (by H. A. Brück)
M. A. Ellison, 1964, p.18 (by E. A. Baker)
W. M. H. Greaves, 1957, p.20 (by W. M. Smart)

Publications issued by the Royal Observatory Edinburgh

Astronomical Observations, Vols I–XIV, 1838–86.
Annals, Vols I–III, 1902–10.
Publications, Vols I–IX, 1955–73.
Communications, Nos 1–208, 1949–75.
Geophysical Communications, Nos 1–7, 1963–67.
Satellite Tracking Reports, Nos 1–200, 1960–75.
Site Testing Reports, Nos 1–72, 1972–75.

Annual Reports

Reports for 17 years between 1846 and 1875, by C. Piazzi Smyth.
Annual Reports for the years 1891–1905, by Ralph Copeland.
Annual Reports for the years 1906–1910, by F. W. Dyson.
Annual Reports for the years 1911–1937, by R. A. Sampson.
Annual Reports for the years 1938–1939, by W. M. H. Greaves.
Annual Reports for the years 1959–1973, by H. A. Brück.
Annual Reports for the years 1974–1975, by V. C. Reddish.